Investigations
Life Science

cpo science

School Specialty Science

CPO Science Life Science Investigations
First Edition
Copyright © 2017 CPO Science, a member of School Specialty Science

ISBN: 978-1-62571-831-0
Part Number 1576075

Printing 2—May 2017
Printed by LSC Communications, US, LLC Kendallville, IN

All rights reserved. No part of this work may be reproduced or transmitted in any form or by any means, electronic or mechanical, including photocopying and recording, or by any information storage or retrieval system, without permission in writing. For permission and other rights under this copyright, please contact:

CPO Science

80 Northwest Boulevard

Nashua, NH 03063

(800) 932-5227

http://www. cposcience.com

Printed and Bound in the United States of America

CPO Science Staff

Scott Eddleman - Author

B.S., Biology, Southern Illinois University; M.Ed., Harvard University

Taught for 13 years in urban and rural settings. Developed two successful science-based school-to-career programs. Nationally recognized teacher trainer in inquiry-based and project-based instruction. Participated in a fellowship at Brown University where he conducted research on the coral reefs of Belize. Worked on National Science Foundation-funded projects at TERC. Scott has been a principal writer and curriculum developer for CPO Science since 1999.

Mary Jo Carabatsos Ph.D. - Principal Writer and Content Editor

Ph.D., in Cell, Molecular and Developmental Biology from Tufts School of Medicine

Taught High School Biology and Physical Science for the last 4 years. Prior to teaching, she worked as a research scientist. Mary Jo advises, edits material and writes investigations.

Erik Benton - Principal Investigation Editor

B.F.A., University of Massachusetts

Taught for 8 years in public and private schools, focusing on inquiry and experiential learning.

Kristen Dolcimascolo - Principal Writer

B.A. Biology from the University of Delaware

Taught middle school science for the past 6 years. She has also taught middle school math. Currently teaching 7th grade at Wayland Middle School in Massachusetts.

Tom Hsu, Ph.D - Principal Writer

Ph.D. Massachusetts Institute of Technology

Nationally recognized innovator in science and math education and the founder of CPO Science.

Senior Editor

Lynda Pennell - Executive Vice President

B.A., English, M.Ed., Administration, Reading Disabilities, Northeastern University; CAGS Media, University of Massachusetts, Boston

Nationally known in high school restructuring and for integrating academic and career education. Served as the director of an urban school for 5 years and has 17 years teaching/administrative experience. In the Boston Public Schools. Lynda has led the development for CPO Science since 1999.

Editorial Consultants

Christine Golden

Chris has been the Project Manger for *Imperial Communications* since 1999 and in the publishing business for 22 years. She is now managing editor and owner of *Big Dog Publishing* Services. Christine's work centers around the editing of K-12 textbook materials.

Contributors

Patricia Davidson - Writer

BA in Biology from Bowdoin College, Brunswick ME, M.Ed., University of New England, Biddeford ME

Teaches seventh-grade life science. Has been teaching for eight years and also teaches nature classes in New Hampshire.

Kelly A. Story - Reviewer

B.S., in Chemistry and English from Gordon College and a Masters in Chemistry from the University of Massachusetts Lowell

Kelly has taught chemistry and is currently a Lab Instructor at Gordon College, MA.

Laine Ives - Writer

B.A., from Gordon College and graduate coursework at Cornell University's Shoals Marine Laboratory and Wheelock College

Laine has taught high school English overseas and environmental education at a middle school in New England.

Jill Elenbaas - Reviewer

B.A., in Biology and Environmental Science from Bowdoin College in Maine

She is currently teaching eighth grade Earth Science in Wayland, MA and has teaching experience as a seventh grade Life Science teacher in Dedham, MA.

Mary Beth Abel Hughes – Writer

B.S., Marine Biology, College of Charleston; M.S., Biological Sciences, University of Rhode Island

Mary Beth taught science and math at an innovative high school. She has expertise in scientific research and inquiry-based teaching methods. She has been a principal writer at CPO since 2000.

REVIEWERS

Pamella Ferris
Physics Teacher
Evans High School, Evans, GA

Brian E. Goodrow
Physical science teacher
Apple Valley Middle School, Apple Valley, CA

Sylvia Gutman
Science teacher, department chairwoman
David A. Brown Middle School
Wildomar, CA Lake Elsinore Unified School District

Tony Heinzman
Science teacher
Apple Valley Middle School, Apple Valley, CA

Philip L. Hunter
Science department chairman
Johnson Middle School, Westminster, CA

Nancy Joplin
English-language arts department chairwoman
Ray Wiltsey Middle School, Ontario, CA

Brad Joplin
Science teacher
Ray Wiltsey Middle School, Ontario, CA

Margaret J. Kilroy
Chemistry Teacher
Cartersville High School, Cartersville, GA

Dakhine Lee
Special Education Department Chair
Lewis Fraiser Middle School, Hinesville, GA

Jason C. Lee
Science Teacher
Long County High School, Ludowici, GA

Mark Matthews
Physics and Chemistry Teacher
Union Grove High School, McDonough, GA

Kelly McAllister
Science teacher
Gage Middle School, Riverside, CA

Bianca N. McRae
Science teacher, department chairwoman
Menifee Valley Middle School, Menifee, CA

Jodye Selco, Ph.D.
Professor, Center for Education and Equity in Math, Science, and Technology
California State Polytechnic University, Pomona, CA

Tia L. Shields
Life science/health and English language learning teacher
Nicolas Junior High School, Fullerton, CA

Sharon Strefling
Science Teacher
Camden County High School, Kingsland, GA

Robert M. Strong
6th Grade Language Arts Teacher
Union Grove Middle School, McDonough, GA

ILLUSTRATION/DESIGN

Polly Crisman -Designer and Illustrator
B.F.A., University of New Hampshire
Graphic artist with expertise in advertising and marketing design, freelance illustrating, and caricature art.

Bruce Holloway - Cover Designs
Pratt Institute, N.Y.; Boston Museum School of Fine Arts
Expertise in product design, advertising, and three-dimensional exhibit design. Commissioned for the New Hampshire Duck Stamp for 1999 and 2003.

Jesse Van Valkenburgh - Designer and Illustrator
Graduated from the Rochester Institute of Technology with a B.A. in Illustration
Jesse has worked at PC Connection as a graphic designer for catalog and direct mailing design, logo design, as well as some illustration work. He has also has experience in creative design and film production.

James Travers - Graphic designer and animator
Associate's Degree of Applied Business and Commercial Art, Akron University
Jim has held positions as graphic designer, art development manager, and currently is a commissioned artist.

EQUIPMENT DESIGN

Thomas Narro - Senior Vice President
B.S., Mechanical engineering, Rensselaer Polytechnic Institute
Accomplished design and manufacturing engineer; experienced consultant in corporate reengineering and industrial-environmental acoustics.

Danielle Dzurik
B.S., Industrial Design, Auburn University
Danielle is responsible for helping to develop new products and improve upon older designs.

Thomas C. Hsu, Ph.D
Ph.D. in Applied Plasma Physics from the Massachusetts Institute of Technology
He is a nationally recognized innovator in science and math education and the founder of CPO Science. Well known as a consultant, workshop leader and developer of curriculum and equipment for inquiry based learning in science and math.

MATERIAL SUPPORT

Kathryn Gavin - Purchasing and Quality Control Manager
Kathryn is responsible for all functions related to purchasing raw materials and quality control of finished goods. She works closely with product development and design.

Lisa LaChance - Senior Materials Specialist
Associate's of Science in Accounting
Lisa evaluates material samples to ensure materials meet project requirements. She develops and manages the release of materials specifications.

TECHNICAL SUPPORT

Tracy Morrow
B.A., English, Texas A&M University; M.A., English, Sam Houston State University, TX
Tracy taught middle school in Klein ISD, a suburb of Houston, Texas, for 9 years preparatory and English at Tomball College for 5 years. Tracy worked as a technical writer in the oil and gas, airlines, and real estate industries. Currently she offers consulting services and technical training. Tracy's expertise lies in the editing program Framemaker.

How to Read an Investigation

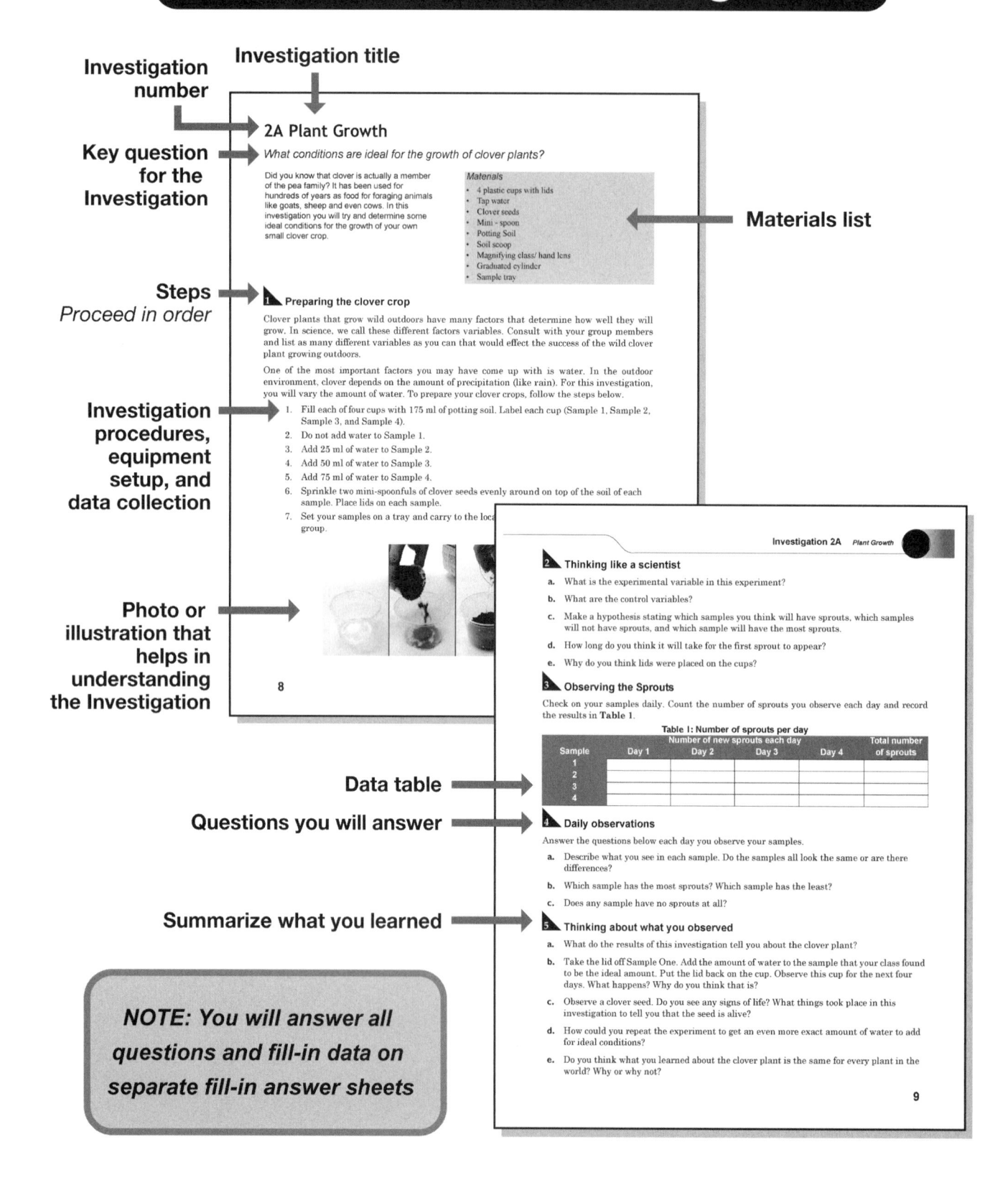

Investigation number → 2A

Investigation title → Plant Growth

Key question for the Investigation → *What conditions are ideal for the growth of clover plants?*

Did you know that clover is actually a member of the pea family? It has been used for hundreds of years as food for foraging animals like goats, sheep and even cows. In this investigation you will try and determine some ideal conditions for the growth of your own small clover crop.

Materials list →

Materials
- 4 plastic cups with lids
- Tap water
- Clover seeds
- Mini - spoon
- Potting Soil
- Soil scoop
- Magnifying glass/ hand lens
- Graduated cylinder
- Sample tray

Steps *Proceed in order* →

1. Preparing the clover crop

Clover plants that grow wild outdoors have many factors that determine how well they will grow. In science, we call these different factors variables. Consult with your group members and list as many different variables as you can that would effect the success of the wild clover plant growing outdoors.

One of the most important factors you may have come up with is water. In the outdoor environment, clover depends on the amount of precipitation (like rain). For this investigation, you will vary the amount of water. To prepare your clover crops, follow the steps below.

Investigation procedures, equipment setup, and data collection →

1. Fill each of four cups with 175 ml of potting soil. Label each cup (Sample 1, Sample 2, Sample 3, and Sample 4).
2. Do not add water to Sample 1.
3. Add 25 ml of water to Sample 2.
4. Add 50 ml of water to Sample 3.
5. Add 75 ml of water to Sample 4.
6. Sprinkle two mini-spoonfuls of clover seeds evenly around on top of the soil of each sample. Place lids on each sample.
7. Set your samples on a tray and carry to the loc... group.

Photo or illustration that helps in understanding the Investigation →

Investigation 2A *Plant Growth*

2. Thinking like a scientist

a. What is the experimental variable in this experiment?
b. What are the control variables?
c. Make a hypothesis stating which samples you think will have sprouts, which samples will not have sprouts, and which sample will have the most sprouts.
d. How long do you think it will take for the first sprout to appear?
e. Why do you think lids were placed on the cups?

3. Observing the Sprouts

Check on your samples daily. Count the number of sprouts you observe each day and record the results in **Table 1**.

Data table →

Table 1: Number of sprouts per day

Sample	Day 1	Day 2	Day 3	Day 4	Total number of sprouts
1					
2					
3					
4					

Questions you will answer →

4. Daily observations

Answer the questions below each day you observe your samples.

a. Describe what you see in each sample. Do the samples all look the same or are there differences?
b. Which sample has the most sprouts? Which sample has the least?
c. Does any sample have no sprouts at all?

Summarize what you learned →

5. Thinking about what you observed

a. What do the results of this investigation tell you about the clover plant?
b. Take the lid off Sample One. Add the amount of water to the sample that your class found to be the ideal amount. Put the lid back on the cup. Observe this cup for the next four days. What happens? Why do you think that is?
c. Observe a clover seed. Do you see any signs of life? What things took place in this investigation to tell you that the seed is alive?
d. How could you repeat the experiment to get an even more exact amount of water to add for ideal conditions?
e. Do you think what you learned about the clover plant is the same for every plant in the world? Why or why not?

NOTE: You will answer all questions and fill-in data on separate fill-in answer sheets

Science Safety Practices and Procedures

❶ Read and follow the Science safety guidelines for each investigation, activity, or field experiment.

1. **Listen** to all teacher instructions before, during, and after investigations.
2. **Prepare** for each investigation or activity.
   a. **Sign** the Science Safety Student Responsibility Agreement.
   b. **Read** each activity or investigation carefully.
   c. **Identify** the investigation purpose.
   d. **Work** ONLY on activities approved by your teacher.
   e. **Follow** all oral and written safety instructions.
   f. **Know** the location of Emergency Safety Equipment such as fire extinguisher, eye and face wash station, safety shower, and first aid kit.
3. **Dress** for laboratory work.
   a. **Wear** protective equipment such as chemical splash goggles, laboratory aprons, and protective gloves as needed.
   b. **Roll** long sleeves above the wrist.
   c. **Tie** back long hair.
   d. **Remove** dangling jewelry and any loose or bulky outer layers of clothing.
   e. **Wear** shoes that enclose the feet. (no flip flops, sandals or open-toe shoes).
4. **Prevent** unsafe situations.
   a. **Be aware** of classmates and their safety.
   b. **Do not** touch, taste or smell any substance without teacher instructions.
   c. **Never** work alone in the laboratory.
   d. **Don't Enter** Science or chemical storage or preparatory areas without a teacher.
   e. **Keep** your work area clean and uncluttered.
5. **Dispose** of used or unused materials according to teacher directions.
6. **Return** equipment to the proper location.
7. **Wash** hands with soap and warm water for 20 seconds after experimenting.
8. **Act** wisely in an emergency.
   a. **Notify** your teacher of any accident immediately.
   b. **Follow** emergency procedures in event of accident.

Science Safety Practices and Procedures (cont.)

❷ Know what to do when...

1. **working with glassware.**
    a. **Don't use** glassware that is chipped or cracked.
    b. **Use special care** to prevent breakage and cuts or scratches.

2. **working with heat.**
    a. **Wear** heat-resistant gloves at all times.
    b. **Do not** touch hot items with bare hands. Use heat resistant gloves, pads or tongs.
    c. **Heat** water only in open containers of heat-resistant glass.
    d. **Watch** all burners, hot plates or open flames.
    e. **Warn** others if they come close to your hot items or liquids.

3. **working with electricity.**
    a. **Keep** electric cords away from water.
    b. **Don't use** frayed cords or plugs in outlets.

4. **finished experimenting.**
    a. **Return** clean materials to their proper locations.
    b. **Dispose** of all used solids and liquids according to teacher instructions. Do not put items in trash or wash down sink without permission.
    c. **Wash** your hands with soap and water for 20 seconds.

5. **you have safety concerns. Tell your teacher, or get help immediately if:**
    a. **You have** trouble using your equipment.
    b. **You do not** understand the instructions for the investigation.
    c. **You injure** yourself, or see someone injured.
    d. **You see** or smell something burning.
    e. **You smell** chemical or gas fumes.

Lab Safety

	General safety: Follow all instructions and safety rules to avoid injury to yourself or others.
	Wear safety goggles: Requires you to wear protective eyewear such as chemical splash goggles or safety glasses to prevent eye injuries.
	Wear a lab apron or coat: Requires you to wear a lab apron or coat to prevent damage to clothing and to protect from possible spills.
	Wear gloves: Requires you to protect your hands from injury due to heat or chemicals.
	Hazardous chemicals: Requires you to use extreme caution when working with chemicals in the laboratory and to follow all safety and disposal instructions from your teacher.
	Skin irritant: Requires you to use extreme caution when handling chemicals in the laboratory due to possible skin irritation and to follow all safety and disposal instructions from your teacher.
	Respiratory irritant: Requires you to perform the experiment under a laboratory hood and to avoid inhaling fumes while handling the chemicals.
	Laser: Requires you to use extreme caution while using a laser during investigations and to follow all safety instructions.
	Sharps: Requires you to use extreme caution when handling sharp objects such as scalpels or glass.
	Electrical: Use caution when working with electricity.

Table of Contents

1A	Measurement and Data	1
1B	Variables in an Experiment	4
2A	Plant Growth	7
2B	Brine Shrimp	10
3A	Classifying Living Things	13
3B	Dichotomous Keys	16
4A	Carbon Dioxide and Living Things	19
4B	Variables and Growth	23
5A	Energy and Ecosystems	26
5B	Testing Pollutants	28
6A	Environments Around School	31
6B	Field Study	35
7A	Examining Onion Tissue	38
7B	Animal and Plant Cells	40
8A	Diffusion and Osmosis	43
8B	Photosynthesis and Color	46
9A	Protozoans	48
9B	Investigating Pond Water	50
10A	Observing the Cell Cycle	53
10B	Modeling Mitosis and Meiosis	55
11A	Observing Human Traits	59
11B	Crazy Traits	62
12A	The DNA Molecule	66
12B	DNA Forensics	69
13A	Crazy Adaptations	71
13B	Natural Selection	74
14A	Relative Dating	77
14B	Interpreting The Fossil Record	81
15A	Creature Cladogram	88
15B	Bread Mold	91

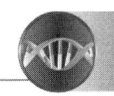

16A	Leaf Structure and Function	93
16B	Flower Dissection	95
17A	Observing Planarians	99
17B	The Mammalian Eye	102
18A	Who's Got the Beat?	106
18B	The Pressure's On	109
19A	Levers	112
19B	Levers and the Human Body	114
20A	Color Vision	116
20B	Light and Vision	120

Additional Materials

Safety Skills	126
Writing a Lab Report	132
Measuring Length	135
Measuring Temperature	139
Calculating Volume	141
Measuring Volume	144
Measuring Mass with a Triple Beam Balance	146
Using a Compound Microscope	149
Recording Observations in the Lab	154
Explore Further: Measuring Biodiversity	158
Explore Further: Innovation and Recycling	161

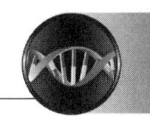

1A Measurement and Data

Is there a relationship between human wingspan and height?

Birds have a wingspan. To measure it, scientists spread the bird's wings and measure from the tip of one wing to the tip of the other wing.

Materials
- Metric tape measure
- Pencil
- Graph paper

You also have a wingspan. Stretch your arms apart at your sides. Your wingspan is the measurement from fingertip to fingertip. In this investigation, you will look for a relationship between human wingspan and height. In the process, you will learn to make accurate measurements.

1. Measuring wingspan and height

1. You will work with a lab partner. Have your partner spread his/her arms out straight from the shoulders on each side of the body as shown below.
2. *Estimate* your partner's wingspan, in centimeters. Record your estimate in Table 1.
3. Using a metric tape measure, measure the distance in centimeters from the tip of the longest finger on one of your partner's hands to the tip of the longest finger on the other hand. This is the wingspan. Your measurement should be to the nearest millimeter (the smallest divisions on the tape measure). Enter the measurement, in centimeters, in Table 1.
4. Have your partner remove her/his shoes. *Estimate* your partner's height in centimeters. Record your estimate in Table 1.
5. Measure your partner's height from the bottom of the heel to the top of the head. Record the data in Table 1.
6. Calculate the difference between your estimate and the actual measurement and record your values in Table 1.

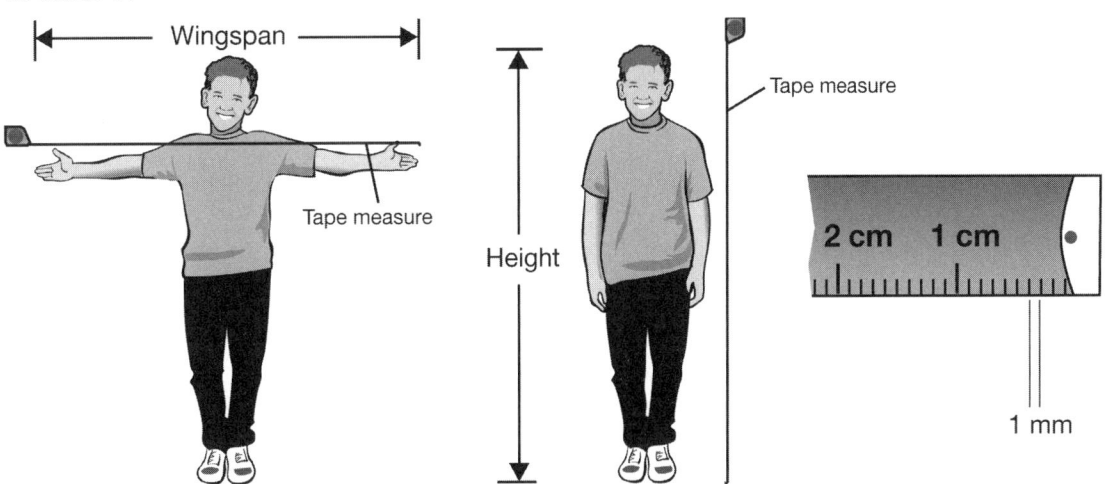

Table 1: Wingspan and height for you and your partner

	Wingspan (*estimate* - cm)	Wingspan (actual - cm)	Difference (larger value minus smaller value)
Lab partner			
You			
	Height (*estimate* - cm)	Height (actual - cm)	Difference (larger value minus smaller value)
Lab partner			
You			

2. Arguing from evidence

a. Which do you think is a better unit for measuring wingspan and height: centimeters or meters?

b. Look at the difference between your estimate and the actual measurements (the last column in Table 1). Did the difference decrease as you took more measurements? Explain these results.

c. Based on the results of your measurements, do you think there is a relationship between human wingspan and height? You will state your answer in the form of a hypothesis. A good hypothesis contains two parts. "If __[I do this]__, then __[this]__ will happen." Your teacher will help you plan how to test your hypothesis in Part 3.

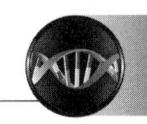

Investigation 1A   *Measurement and Data*

3. Gathering class data

1. Record your data (actual measurements only) on the board in front of the classroom.
2. Once the class data table is complete, copy the data into Table 2.
3. Complete Table 2 by finding the difference between wingspan and height for each row. Subtract the lesser amount from the greater amount to find the difference.
4. Look at the data in Table 2. Do the data support your hypothesis? Explain.
5. After reading Section 1.3 in your text, make a graph of the data in Table 2. Plot wingspan on the *x*-axis and height on the *y*-axis. Can you identify a relationship on the graph?

Table 2: Class data

Student	Wingspan (cm)	Height (cm)	Difference

4. On your own

a. Take measurements of wingspan and height for 10 people outside of your classroom. Create your own data table and graph.

b. Is your data consistent with your class data? What can you conclude from your results?

1B Variables in an Experiment

How do scientists conduct a good experiment?

Imagine a jumping frog trying to escape from a predator. The frog needs to get the greatest distance out of each jump. What variables affect how far the frog will travel? What about the angle at which the frog aims when it jumps? How do you think the launch angle will affect the distance the frog will travel? In this Investigation, you will try launching marbles (not frogs!) at different angles in order to find out how launch angle affects distance traveled. As a result, you'll learn how to conduct a good experiment.

Materials
- Marble launcher
- Plastic marbles
- Tape
- Metric tape measure
- Graph paper
- Ruler
- Safety goggles

Safety Note:
- **Never launch marbles at people.**
- **Wear safety goggles or other eye protection when launching marbles.**
- **Launch only the plastic marbles that come with the marble launcher.**

1. Setting up

1. Identify the parts of the marble launcher.
2. For this experiment, you will use the fifth slot of the barrel to launch the marble for each trial. Pull the launch lever back and slip it sideways into the fifth slot. Put a marble in the end of the barrel. The marble launcher is now ready to launch.
3. You will change the angle for each launch starting at 10 degrees and increasing 5 degrees up to 80 degrees.
4. A minimum of two people are needed per launcher. One person releases the launch lever and the other watches where the marble lands. A few launches should be done at each angle to be sure that the data is accurate. It also takes a few times to accurately find the spot where the marble lands.
5. Use a strip of masking tape on the floor to make sure that the marble launcher is set back in the same place every time. A tape measure laid along the floor provides a good way to measure the distance traveled by the marble.

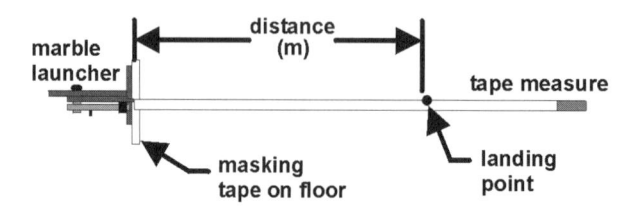

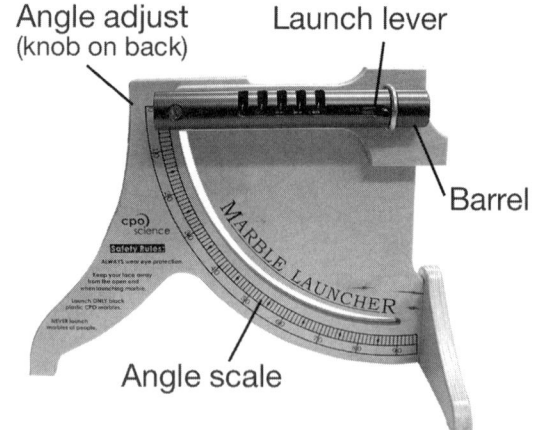

Investigation 1B Variables in an Experiment

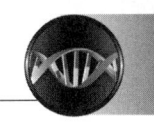

2. Stop and think

a. At which angle do you think the marble will travel the greatest distance? State your answer to the question as a _hypothesis_.

b. What is the _experimental variable_ in this experiment? What are the _control variables_?

c. Why is it important to make a few practice launches at each angle?

3. Doing the experiment

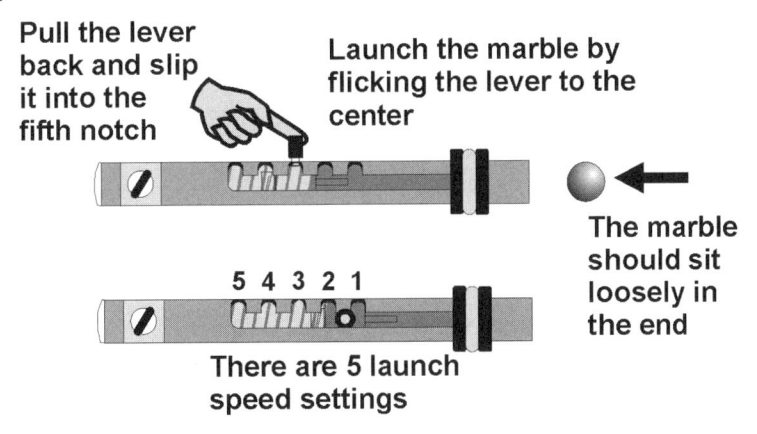

Pull the lever back and slip it into the fifth notch

Launch the marble by flicking the lever to the center

The marble should sit loosely in the end

5 4 3 2 1

There are 5 launch speed settings

In the table below, record your two best trials for each launch angle.

Table 1: Launch angle and distance data

Launch angle (degrees)	Distance (meters)	Distance (meters)	Launch angle (degrees)	Distance (meters)	Distance (meters)
10			50		
15			55		
20			60		
25			65		
30			70		
35			75		
40			80		
45			85		

4. Analyzing your data

a. There is an angle at which the marble launcher will cause the marble to travel the farthest. The angle may not be obvious from the data you have collected. Graphs help scientists to organize data into patterns that are easier to see. For graphing purposes, which variable is the *independent variable*? Which is the *dependent variable*?

b. Make a line graph showing how the distance changes with the launch angle. Plot the independent variable on the *x*-axis and dependent variable on the *y*-axis.

c. Look at your graph. At what angle does the marble attain the greatest distance?

d. You are challenged to launch a marble to travel a distance of 4.00 meters. At what angle will you set the launcher?

e. Referring to your answer for question (d), state another angle that would give you the same result.

f. Is the 4.00 meter distance the only distance that you can reach using two different angles? State three other distances and the angles you would use to reach that distance.

g. Explain why two angles can be used to reach the same distance.

h. Write a paragraph about a situation in which it would be better to use one angle rather than the other.

5. Designing your own experiment

a. Besides launch angle, which other variables can you change on the marble launcher?

b. Choose a different variable you can change. Write a question you have about how that variable affects another variable.

c. State your hypothesis to the question.

d. Design an experiment to test your hypothesis. List the materials and procedures for your experiment, then get approval from your teacher.

e. Conduct your experiment. Be sure to make a good data table before you begin. Get approval from your teacher for your data table design.

f. Make a graph of your data and analyze your results.

g. State a conclusion to your experiment. Did your results support your hypothesis? If so, explain why. If not, explain how you would change your hypothesis or experiment.

h. Present your experiment and results to the class.

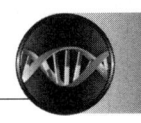

2A Plant Growth

What conditions are ideal for the growth of clover plants?

Did you know that clover is actually a member of the pea family? It has been used for hundreds of years as food for foraging animals like goats, sheep, and even cows. In this investigation, you will try to determine some ideal conditions for the growth of your own small clover crop.

Materials

- 4 Plastic cups with lids
- 250-mL graduated beaker
- Potting soil
- 100-mL graduated cylinder
- Tap water
- Mini spoon or straw cut at tip
- Clover seeds
- Plastic tray for samples
- Magnifying glass/hand lens

1. Preparing the clover crop

Clover plants that grow wild outdoors have many factors that determine how well they will grow. In science, we call these different factors *variables*. Consult with your group members and list as many different variables as you can that would affect the success of the wild clover plant growing outdoors.

One of the most important variables you may have come up with is water. In the outdoor environment, clover growth depends on the amount of precipitation (like rain). For this investigation, you will vary the amount of water. To prepare your clover crops, follow these steps:

1. Fill each of four cups with 175 mL of potting soil. Label each cup (Sample 1, Sample 2, Sample 3, and Sample 4).
2. Do not add water to Sample 1.
3. Add 25 mL of water to Sample 2.
4. Add 50 mL of water to Sample 3.
5. Add 75 mL of water to Sample 4.
6. Sprinkle two mini-spoonfuls of clover seeds evenly around on top of the soil of each sample. Place lids on each sample.
7. Set your samples on a tray, and carry to the location designated by your teacher.

2. Thinking like a scientist

a. What is the <u>experimental variable</u> in this experiment?

b. What are the <u>control variables</u>?

c. Complete the hypothesis that begins: If _____ determines the total number of sprouts per day, then the number of sprouts will be _____ in Sample 1, _____ in Sample 2, _____ in Sample 3, and _____ in Sample 4.

d. How long do you think it will take for the first sprout to appear?

e. Why do you think lids were placed on the cups?

3. Observing the sprouts

Check on your samples daily. Count the number of sprouts you observe each day and record the results in Table 1.

Table 1: Number of sprouts per day

Sample	Day 1	Day 2	Day 3	Day 4	Total number of sprouts
1					
2					
3					
4					

(Number of new sprouts each day)

4. Daily observations

Answer the questions below each day you observe your samples.

a. Describe what you see in each sample. Do the samples all look the same or are there differences?

b. Which sample has the most sprouts? Which sample has the fewest?

c. Does any sample have no sprouts at all?

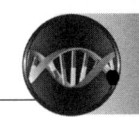

5. Communicating your results

a. What do the results tell you about the clover plant? Share your conclusion with the class.

b. Take the lid off Sample 1. Add the amount of water to the sample that your class found to be the ideal amount. Put the lid back on the cup. Observe this cup for the next four days. What happens? Why do you think that is?

c. Observe a clover seed. Do you see any signs of life? What things took place in this investigation to tell you that the seed is alive?

d. How could you repeat the experiment to get an even more exact amount of water to add for ideal conditions?

e. Do you think what you learned about the clover plant is the same for every plant in the world? Why or why not?

2B Brine Shrimp

What conditions are ideal for the growth of brine shrimp?

Brine shrimp are crustaceans that live in salty bodies of water. These include the Great Salt Lake in Utah, estuaries, and brackish ponds. These different bodies of water can have different amounts of salt in them. In the first part of this investigation, you will hatch some brine shrimp in four different samples of water. Each sample will have a different amount of salt in it. Over the course of the next few days, you will check your four samples of water to see how many brine shrimp have hatched. They are very small, but you will be able to see them and even count them if you look very closely.

Materials

- 4 Plastic cups with lids
- Aged tap water (2 liters)
- Brine shrimp eggs
- Mini-spoon or cut straw tip
- Sea salt or Kosher salt
- Salt scoop
- Magnifying glass/hand lens
- Digital or compound microscope (400x)
- 100-mL graduated cylinder
- 250-mL beaker
- Sample tray

WARNING — This lab contains chemicals that may be harmful if misused. Read cautions on individual containers carefully. Not to be used by children except under adult supervision.

1. Preparing the samples

1. From your 2 liters of water, measure out 150 mL of water into the beaker.
2. Fill one of your four plastic cups up with the water. Mark this cup "Sample 1."
3. Fill and label the other three samples with 150 mL of water each.

Now that you have four samples of water, you need to decide how much salt to put into each sample. One sample should contain no salt at all. Many times scientists change the variables in their experiments in even amounts. Decide how many scoops of salt you want to increase each sample by, like two scoops each time. For example:

- Sample 1 - zero scoops
- Sample 2 - two scoops
- Sample 3 - four scoops
- Sample 4 - six scoops

Use level scoops when you add salt to a sample. Label how many scoops went into each sample on the lid of each cup. Once the samples have been labeled, calculate the parts per thousand concentration of salt in each. To do this, follow the directions on the next page.

Investigation 2B *Brine Shrimp*

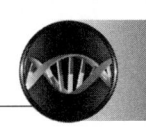

Measure the mass of one scoop with a balance. The number of scoops in your sample times the mass of one scoop of salt is the mass of salt in each sample. Calculate the number of grams of salt for each sample and record the information in Table 1 to calculate the parts per thousand of salt in each sample. There are 1,000 grams of water in a liter, so to find the parts per thousand of salt in the water,

$$\frac{\text{number of grams of salt}}{150 \text{ mL of water}} \times \frac{1,000 \text{ mL of water}}{1,000 \text{ grams of water}} = \frac{\text{number of grams of salt}}{1,000 \text{ grams of water}}$$

Table 1: Salt concentration

Sample number	Grams of salt (no. of scoops x mass of 1 scoop)	Parts per thousand

2. Thinking like a scientist

a. What is the experimental variable in this experiment?

b. What are the control variables?

c. What do you think will happen when the eggs are added to the samples?

d. Which sample do you think will have the most shrimp in two days?

e. Which sample do you think will have the fewest shrimp in two days?

f. Do you think the sample with no scoops of salt added will have any shrimp in two days? Why or why not?

3. Hatching the shrimp

Now that you have your samples prepared, it is time to add the shrimp eggs. Each sample should have the same size mini-spoonful of eggs in it.

1. Add one mini-spoonful to each sample. (Even though it is very small, one mini-spoonful is plenty of eggs to hatch a whole colony in each sample!)
2. Put the cover on each sample, and set the cups into the sample tray.
3. Make sure all samples are properly labeled and capped.
4. Store the samples in a place that will have light for at least part of the day, like a classroom.

11

4 ▶ Observing the results

Carefully look at each of your samples. What do you see?

Record your observations every 24 hours in your notebook.

- a. Describe what you see in each sample.

- b. Do the samples all look the same or are there differences?

- c. Which sample has the most brine shrimp and which has the fewest?

- d. Have you made any other observations?

5 ▶ Communicating your results

- a. What does this part of the investigation tell you about the ideal level of salt the brine shrimp need to thrive?

- b. Why is it useful to include a sample that has no salt added to it?

- c. If some shrimp hatched in a sample of water that had no salt added, what would that tell you about that sample of water?

- d. Compare your results to the results of all the groups in your class. Overall, what seems to be the best salt concentration for hatching brine shrimp?

- e. How could you modify the experiment to get a more precise ideal salt concentration for hatching brine shrimp?

Investigation 3A *Classifying Living Things*

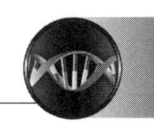

3A Classifying Living Things

How do we the classify living things found around the schoolyard?

Scientists estimate there are somewhere between 5 and 30 million different types of living things on Earth! Each different type of living thing is called a <u>species</u>. How can scientists begin to understand and learn about so many different species? To begin with, they create groups based on observable characteristics. By learning about one organism in the group, they make inferences about others in the same group. This is known as classifying. In this investigation, you will practice classifying organisms you find in your schoolyard. Then you will work with a partner to create a classification system.

Materials

- Clipboard
- Pencil or pen
- Observation sheets or science journal
- Metric ruler or tape measure
- Magnifying lens
- Digital camera
- Other materials supplied by your teacher

Safety Note:
- Dress appropriately when working outside.
- Never place any plant part in your mouth.
- Never rub any sap or fruit juice into your skin or an open wound.
- Never eat food after handling plants without first scrubbing your hands.

1. Think like a scientist

a. What types of organisms do you expect to find around your schoolyard?

b. Where in the schoolyard might you look to increase your chances of finding a variety of different organisms?

c. What are some observable characteristics you can use to compare different organisms?

2. Observing organisms

1. Obtain a clipboard and observation sheets from your teacher.
2. If your teacher provides them, obtain other tools and supplies for making and recording observations.
3. In small groups, examine the schoolyard for living organisms. Be sure to follow your teacher's directions about boundaries.
4. Using the clipboard and observation sheets, record your data about the organisms you find. An example observation sheet is shown on the next page.
5. When you return to the classroom, your teacher may provide other living or preserved organisms for your class to observe and evaluate.

13

3. Sample observation sheet

Directions: Observe each organism. Use the observation sheet to record your observations. For each organism, include as much of the suggested information as you can find. Also, make a careful sketch of the organism in the space provided. If you don't know the common name of the organism, consult other classmates or your teacher. You may use a general common name like "mushroom," "beetle," or "grass."

Sketch | **Information**

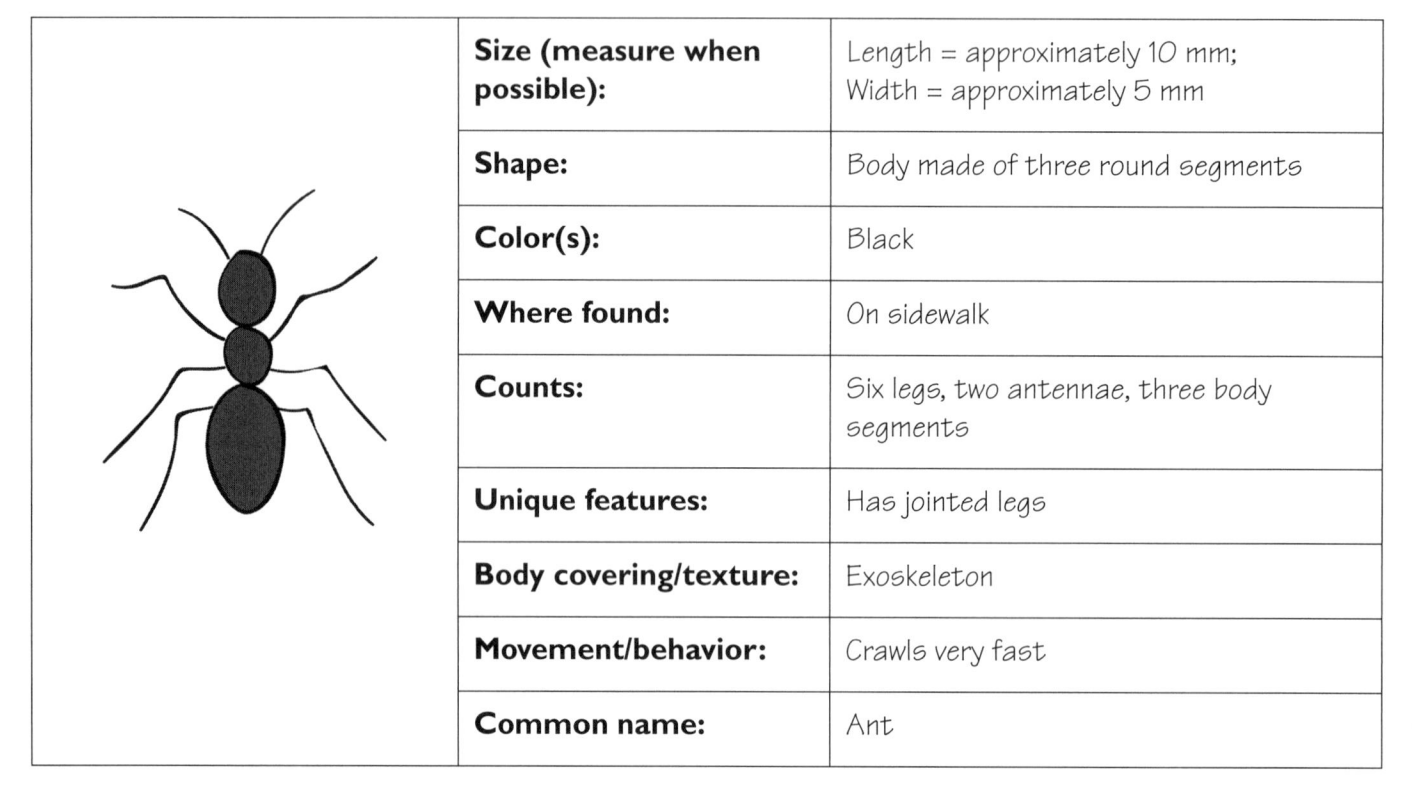

	Size (measure when possible):	Length = approximately 10 mm; Width = approximately 5 mm
	Shape:	Body made of three round segments
	Color(s):	Black
	Where found:	On sidewalk
	Counts:	Six legs, two antennae, three body segments
	Unique features:	Has jointed legs
	Body covering/texture:	Exoskeleton
	Movement/behavior:	Crawls very fast
	Common name:	Ant

4. Classifying organisms

1. Using your textbook, study the four kingdoms and fill in the information in Table 1.

Table 1: Characteristics of the four kingdoms

Kingdom	What do all members of the kingdom have in common?
Protista	
Fungi	
Plantae	
Animalia	

Investigation 3A *Classifying Living Things*

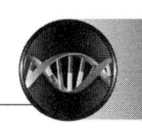

2. Examine your observation sheets. Determine the kingdom to which each organism belongs. List the common name of each organism in the correct column of Table 2.

Table 2: Classifying the organisms you observed

Protists	Fungi	Plants	Animals

5. Thinking about what you observed

a. Which kingdoms were not found in your school yard? Suggest reasons why you may not have found them. What additional tools or types of places would you need to locate organisms from all six kingdoms?

b. Using magazines, the Internet, and other sources, research organisms from missing kingdoms so you have representatives from all six kingdoms.

c. We classify things in many places and for many reasons. For example, at the movie store, rentals are organized according to movie types such as comedy, foreign films, etc. This makes finding movies easier and helps us to determine if we want to see a movie or not. In addition to the scientific field and the movie stores, list at least three other places where people use a classification system. Explain how each classification system is organized.

d. Classifying has many benefits; however, scientists need to be careful when using classification to learn about organisms. Try to think of at least one drawback that might occur if scientists use only classification to judge and learn about organisms.

e. Why could you not observe or record data on organisms from Domain Archaea or Domain Bacteria in the schoolyard?

6. Exploring on your own

Create a guidebook to organisms in your schoolyard. Make sure an observation sheet is filled out for each organism you observed. Then organize the sheets so that each kingdom has a chapter. If desired, add photos and other information about each organism.

3B Dichotomous Keys

How do you create a dichotomous key?

A <u>dichotomous key</u> is made from a series of steps, each consisting of two statements. The statements describe the characteristics of an organism or group of organisms. Usually, a dichotomous key starts out with broad characteristics that become more specific as more choices are made. As you read each step, you choose one of the two statements based on the organism's characteristics. Eventually, the statements lead you to the name of the organism or the group to which it belongs. In this investigation you will create a dichotomous key to identify imaginary creatures.

Materials

- Creature Cards
- Key charts
- Paper
- Pencils and pens

1. Setting up

1. Examine the creature cards. Compare and contrast the major features of the creatures.
2. Make up a name for each creature — be imaginative!
3. Make a list in Table 1 of the characteristics that may help you classify the creatures into different groups. Some characteristics to consider include living area (habitat), numbers of appendages, types of body parts, and body covering.

Table 1: Imaginary creature characteristics

Creature number	Creature name	Characteristics
1		
2		
3		
4		
5		
6		
7		
8		
9		
10		

2. Stop and think

a. What types of features are listed in Table 1?

b. Can you think of one feature that would divide all of the creatures into two groups?

Investigation 3B *Dichotomous Keys*

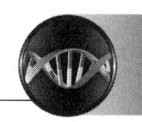

3. Developing a key to the creatures

1. Obtain blank key charts from your teacher.
2. Examine the creatures. Think of a question that, upon answering, separates the creatures into two groups. Write the question into the question box of a key chart and label it Key Chart 1.
3. If the answer to question 1 is YES, write the numbers of the corresponding creatures below the YES box. If the answer is NO, write the numbers of the creatures below the NO box. For the YES group, write *Go to Key Chart* 2 in the space provided.
4. Think of a question that separates the YES group into two more groups. Complete another key chart for the YES group. Continue to complete additional key charts for YES and NO groups until you narrow the original YES group down to individual creatures. Be sure to indicate which key chart number to go to next in the spaces provided. An example key chart is shown below.

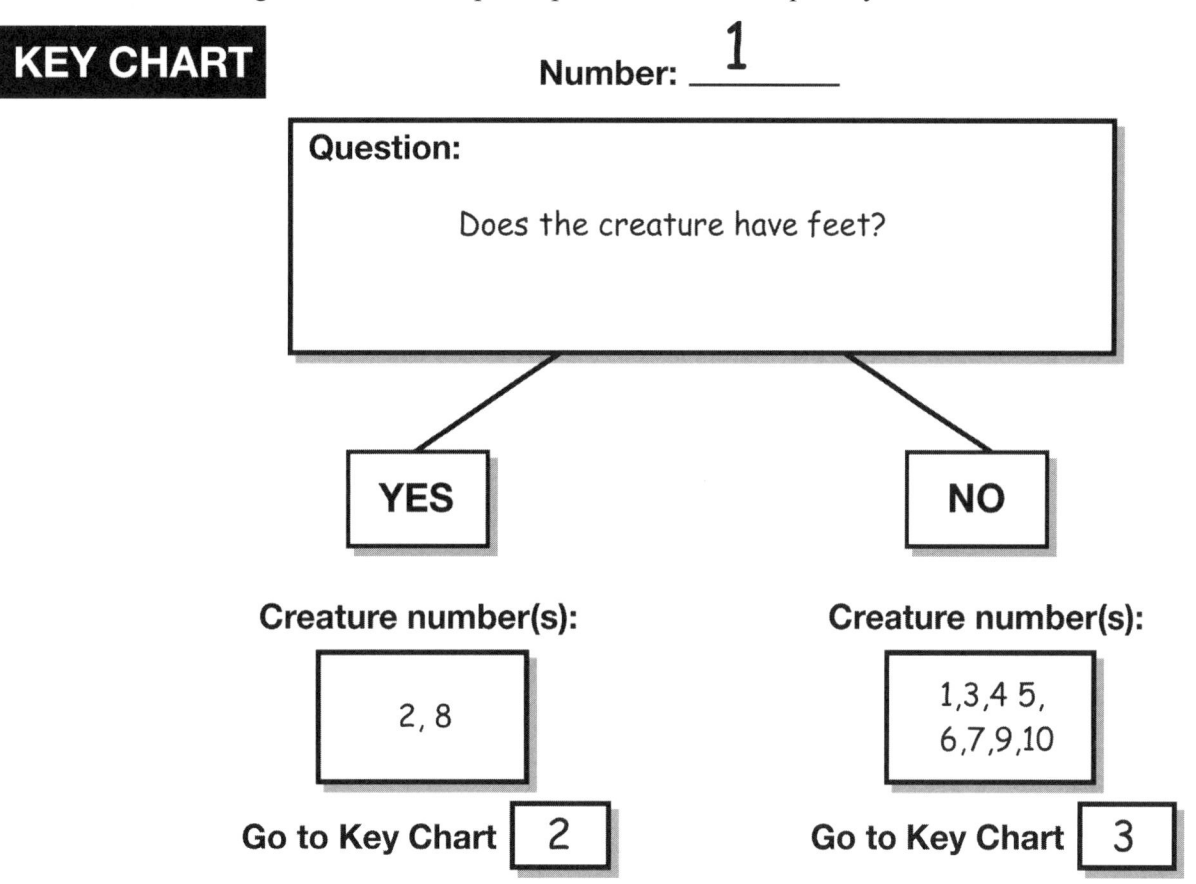

KEY CHART Number: __1__

Question: Does the creature have feet?

	YES	NO
Creature number(s):	2, 8	1,3,4 5, 6,7,9,10
Go to Key Chart	2	3

5. Repeat the process you used in step 4 for the first NO group.
6. When you are finished, you will be able to use your series of key charts to identify each creature.
7. Exchange your charts with another student, and test to make sure someone else can identify your creatures using your key charts.

4. Making a dichotomous key

1. A *dichotomous key* is a tool that helps its user identify natural objects like birds, trees, fungi, and insects. Dichotomous means "divided into two parts." Therefore, dichotomous keys always give you two choices in each step. How are your key charts similar to a dichotomous key? How are they different?

2. A dichotomous key is made from a series of steps, each consisting of two statements. **Use your key charts to create a dichotomous key to identify the creatures.** Use the examples of dichotomous keys in section 3.2 of your textbook as a guide to help you make your own. Your completed dichotomous key should allow you to identify all 10 creatures by the names you gave them.

3. Give your dichotomous key to another classmate and give them one of the creature cards. Have the classmate try to identify the name of that creature using your dichotomous key. If the classmate cannot identify the creature by the name you gave it, you may need to revise your key. Continue to test and revise your key until the creatures can be identified by other classmates.

5. Applying your knowledge

Create a dichotomous key for ten insects, birds, or mammals.

1. Find pictures of ten different insects, birds, or mammals. You may use magazines or the Internet.
2. Place each picture on an index card.
3. Write the name of the organism on the opposite side of the index card. You may use common names or scientific names (if you can find them).
4. Create your dichotomous key.
5. Have a classmate use your key to identify the organisms on your index cards.

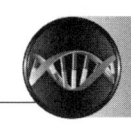

Investigation 4A Carbon Dioxide and Living Things

4A Carbon Dioxide and Living Things

How is carbon dioxide important to living things?

A chemical reaction is a process that rearranges the atoms of one or more substances into one or more new substances. Living cells use many chemical reactions. Plant cells use a chemical reaction called *photosynthesis* to store energy from the sun in the form of molecules. All cells use a chemical reaction called *cellular respiration* to release energy from molecules. Carbon dioxide is a compound that is involved in both reactions. In this investigation, you will use a chemical to determine if carbon dioxide is increasing or decreasing under different conditions.

Materials
- 6 40-mL Vials with screw caps or stoppered test tubes
- Masking tape
- Marker or pen
- Drinking straw
- 250-mL Flask
- 100-mL Graduated cylinder
- 150 mL Bromothymol blue solution (0.04%) in a beaker
- 2 5-cm sprigs of *Elodea* or other aquatic plant
- Funnel
- Aluminum foil
- Forceps
- Seeds: pea or bean
- Light source
- Filter paper cone

Safety Note: Wear safety goggles and an apron during this investigation.

WARNING — This lab contains chemicals that may be harmful if misused. Read cautions on individual containers carefully. Not to be used by children except under adult supervision.

1. Thinking about cellular respiration

In cellular respiration, carbohydrates react with oxygen to produce carbon dioxide and water. Every cell in your body uses this reaction to release energy stored in carbohydrates. When you breathe, you take in oxygen, a reactant in cellular respiration. You exhale carbon dioxide, a product of cellular respiration. All types of cells including animal and plant cells undergo cellular respiration.

Cellular respiration:

Carbohydrate + Oxygen ⟶ Carbon dioxide + Water

1. Pour 100 mL of bromothymol blue solution into a 250-mL flask. Bromothymol blue is a chemical that detects carbon dioxide dissolved in water.

2. Using scissors, cut a small diamond out of the side of a straw, about 2 centimeters from one end. This will allow you to blow through the straw without drawing any chemical solution up into your mouth. Use this straw to GENTLY blow bubbles into the solution until it changes color completely. CAUTION: Avoid getting any of the solution on your face or mouth!

a. What was the initial color of the bromothymol blue solution?

b. What was the color of the solution after you blew bubbles into the flask?

c. What color does bromothymol blue change to when carbon dioxide is present?

d. What do you think would happen to the color of the solution if the carbon dioxide were used up?

19

2. Thinking about photosynthesis

Plant cells use a chemical reaction called *photosynthesis* to store energy from the sun in the form of carbohydrate molecules. In the reaction, carbon dioxide is combined with water to make carbohydrates and oxygen. The word "light" above the arrow means that light is required to make the reaction happen:.

> **Photosynthesis:**
>
> Carbon dioxide + Water $\xrightarrow{\text{Light}}$ Carbohydrate + Oxygen

a. Would the presence of plants in water cause the carbon dioxide to increase, decrease, or stay the same? Explain your answer.

b. When would you expect photosynthesis to happen, during the day or at night? Explain your answer.

3. Investigating photosynthesis—Experiment A

1. Obtain four clear plastic vials with screw caps. Label each vial using a marker and masking tape as shown:
   Vial 1 Plant (uncovered)
   Vial 2 Plant (covered)
   Vial 3 No plant (uncovered)
   Vial 4 No plant (covered)

2. Use the flask of bromothymol blue solution you obtained in Part 1 for this experiment. If needed, gently blow bubbles into the solution again to make sure the color has completely changed. Record the final color of the solution in Table 1.

3. Pour the solution into the graduated cylinder. Using a funnel, fill each vial equally, about 25 mL each. Cap vials 3 and 4.

4. Place a 5-cm piece of aquatic plant in vials 1 and 2. Make sure each plant is completely submerged. Cap these vials.

5. Wrap vials 2 and 4 with aluminum foil to block out all light.

6. Place the four capped vials upright in front of a lamp. Make sure that the vials are at least 20 cm away and that light is hitting all vials equally on one side. Your setup should look similar to the one shown to the right.

7. Let the plants sit for 45 minutes or overnight. Rinse and dry all labware.

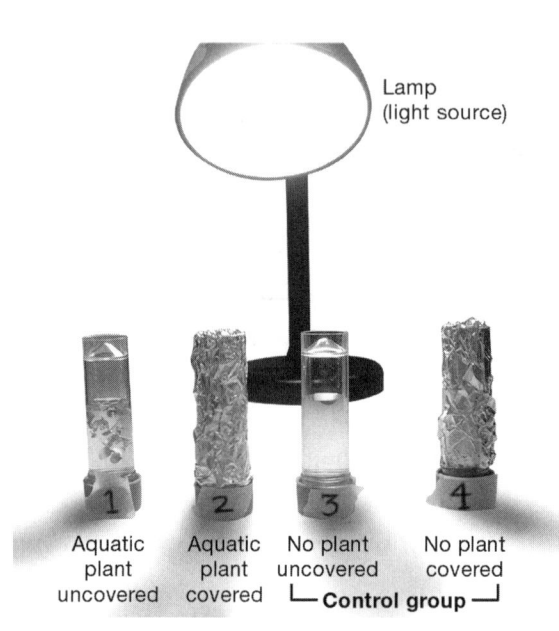

4. Investigating cellular respiration—Experiment B

1. Obtain two plastic vials with screw caps and label each with a marker on tape as follows:
   Vial 5 Seeds and Vial 6 No Seeds.

2. Half-fill Vial 1 with seeds.

Investigation 4A   *Carbon Dioxide and Living Things*

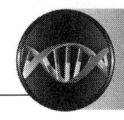

3. To each vial, add 25 mL of fresh bromothymol blue solution. Be sure to use enough solution to cover the seeds. *The initial color of the solution should be blue for this experiment.*
4. Record the initial color of the solution for both vials in Table 2.
5. Cover both vials with foil. Place them with the other vials for at least 45 minutes or overnight
6. Rinse and dry all labware. Dispose of materials according to teacher instructions, clean your area, and wash your hands.

5. Making predictions

a. Why did you need to make sure the bromothymol blue solution had completely changed color before starting Experiment A?

b. In Experiment A, what is the experimental variable? List the variables you tried to control.

c. Predict what you think will happen to the color of the bromothymol blue solution in each of the six vials. Write your predictions in Table 1 and Table 2.

6. Recording data

1. After 45 minutes, or on day 2, remove the foil from each vial.
2. Using tweezers, carefully remove the plants from vial 1 and 2, rinse, and return them.
3. Record the final color of the solutions in vials 1-4 in Table 1.

Table 1: Data from Experiment A: Photosynthesis

Vial number	Initial color	Predicted color	Final color
1 Plant (light)			
2 Plant (dark)			
3 No plant (light)			
4 No plant (dark)			

4. Set up the funnel and filter paper over a beaker. Carefully pour the solution from vial 5 into the beaker. Dispose of the seeds. Use the funnel to pour the solution back into vial 5.
5. Filter the seeds from vial 6 and return the solution.
6. Compare the colors of the two solutions and record your data in Table 2.

Table 2: Data from Experiment B: Cellular respiration

Vial number	Initial color	Predicted color	Final color
5 Seeds (covered)			
6 No seeds (covered)			

7. When finished, follow all cleanup instructions from your teacher and wash your hands.

7. Arguing from evidence

a. From Experiment A, in which vial(s) did photosynthesis occur? Use your data and knowledge of photosynthesis to explain your answer.

b. Did the results from Experiment A support your prediction from Part 5?

c. From Experiment B, in which vial(s) did cellular respiration occur? Use your data and knowledge of cellular respiration to explain your answer.

d. Did the results from Experiment B support your prediction from Part 5?

e. Why was light a variable in Experiment A but not a variable in Experiment B?

f. Explain the importance of bromothymol blue in both experiments.

8. Planning and conducting your own experiment

Yeast is an organism used for thousands of years to make bread rise. Yeast cells convert the sugars in bread to carbon dioxide and water. The carbon dioxide bubbles in the dough make the bread expand. If kept in a warm place, the reaction speeds up and produces larger bubbles more quickly throughout the loaf.

1. Design a simple controlled experiment to test for evidence of cellular respiration in yeast.
2. Write a hypothesis using "If _____ (I do this), then _____ (this will happen)."
3. Use the following materials: package of bread yeast, packets of sugar, empty 1-liter water or soda bottles, water, balloons.
4. Record your data in the form of a chart or pictures, then communicate your findings to your class.

Investigation 4B *Variables and Growth*

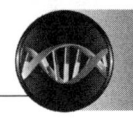

4B Variables and Growth

How will similar populations react to changing variables?

Many variables affect plants as they grow. Clover plants are no exception. Plants have been able to grow in almost every environment on the planet, even under water. Each environment has a wide range of variables that can affect a plant's success and health. In this investigation, you will choose a variable and design an experiment to test how it affects plant growth.

Materials
- 4 Sample clover populations
- 4 Different sized containers
- 4 Lids
- Sand, dirt, and potting soil
- Plant grow light
- Fertilizer
- 100-mL Graduated cylinder
- Beaker
- Measuring spoon/cup

1. Conditions to be tested

Since there are so many variables that influence plants growth, each group in your class will select one to test. Some variables may take longer than others to show a noticeable effect on the clover plants. The list below shows the different variables your group can test.

Direct sunlight | **Partial sunlight** | **Complete shadow** | **Dark**

1. Light levels - direct sunlight, partial sunlight, complete shadow, dark
2. Varying amounts of fertilizer.
3. Different types/brands of fertilizer
4. Different kinds of soil– dirt, sand, or a combination of these
5. Different sized containers

23

Once you have decided with your group which variable you are going to test, prepare your experiment setups. Prepare setups with different levels of the variable you are testing. One of your setups should be the control group. The <u>control group</u> is the experiment that is set up under "normal" conditions. For example, if you are testing light levels, your control group would be the one placed in lighting conditions the plant would normally encounter.

2. Thinking like a scientist

a. What is the experimental variable in your experiment?

b. What are the control variables?

c. What question are you asking with your experiment?

d. Based on your experimental variable, what is your hypothesis? Use the form "If _____ (we do this), then _____ (this will happen)."

e. What is the purpose of testing the effect of varying a condition on both clover plants from seed and existing sprout populations?

f. How will you set up your control group? Explain why it is your control group.

g. How long do you think it will take for the first effects to be observable?

h. Sketch your experiment setup. Label parts.

3. Observing the populations

Now that you have your samples prepared, it is time to look for any effects that the varying conditions may have on your populations. Check on your samples daily and record your group's observations in **Table 1**.

Table 1: Plant population data

Sample	Effect on populations			
	Day 1	Day 2	Day 3	Day 4
1				
2				
3				
4				

4. Observing the results

Carefully look at each of your samples. Try not to disturb the ongoing experiment too much while observing your samples. What do you see?

After 24 hours (on Day 1),

a. Describe what you see in each sample.

b. Do the samples all look the same or are there differences?

c. Which sample seems most affected by the experimental variable? The least affected?

Answer questions a-c for Days 2-4.

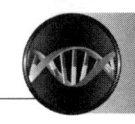

5. Thinking about what you observed

a. What do the results of this investigation tell you about the variable you tested?

b. Do the results support your hypothesis?

c. How could you repeat the experiment to get even more exact data on the effect of your experimental variable?

d. Do you think the variable you tested has the same effect on every other organism on Earth? Why or why not?

e. Can you set up an environment for a clover crop that combines all of the most successful conditions observed by all the groups in your class? Try it out and see what happens.

6. Presenting what you have learned

Work with your group to come up with a presentation that clearly explains what you have learned. All members of the group should be involved in some way. You can use these points as a guideline to make your presentation:

1. State the question your experiment was asking.
2. State your hypothesis.
3. Explain the setup and procedure of your experiment.
4. Give an overview of the data you collected, pointing out any trends or patterns you observed.
5. State your conclusion (does your data support your hypothesis?) and propose an alternative conclusion if one is possible.

5A Energy and Ecosystems

How does energy flow through living systems?

The species in an ecosystem depend on one another in many ways. We can model these relationships using food chains, food webs, and energy pyramids to show the transfer of energy from species to species. Food chains show how each member of an ecosystem gets its food. A food web is a group of overlapping food chains in an ecosystem. An energy pyramid is a diagram that shows how energy moves from one feeding level to the next in a food chain. In this investigation, you will build food chains and food webs as you play a card game!

Materials
- Food Web Builder cards

1. Planning your investigation

1. Your teacher will give you a deck of Food Web Builder cards to look at before you play the game. Spread out the cards to become familiar with each species card.

   - **Name** (scientific and common name of species)
   - **Energy niche** (how the species gets its energy)
   - **Eats** (what the species eats, unless it is a producer)
   - **Eaten by** (if any)

2. Four different ecosystems are shown like suits on playing cards (hearts, diamonds, clubs, or spades). Locate their symbols on the upper corner of each card.

3. Energy can be added or recycled during the game using Sun or Decomposer cards.

2. Developing a model

1. Name the **four** ecosystems in the game.
2. List the **four** energy niches (roles) species can play in an ecosystem.
3. Choose several cards from which to diagram a food chain. Use at least **three** species and be sure to link species with correct arrows.

3. Playing the game

Your teacher will divide your class into groups of four students or less. Here is how to play.

1. The object of the game is to create food chains and food webs from your cards and earn the most points to win.
2. The dealer shuffles all the cards and then deals each student **seven** cards face down. The remaining cards are placed in a "draw pile" in the center of the table. The top card of the draw pile is flipped over to create the "discard pile" next to the draw pile.
3. Students examine and organize cards and see if they have any complete food chains or webs in their hands.
4. Play begins with the student to the left of the dealer. Each turn starts by drawing one card, from the top of either the draw or the discard pile. Each turn ends by placing one card face up on the discard pile. Play continues to the left.

Investigation 5A *Energy and Ecosystems*

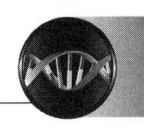

5. During a turn, a student can:
   - lay down a complete food chain of the same suit with at least three species, **OR**
   - add a Sun or Decomposer card to a food chain or web, as long as one is not already in the same food chain or web, **OR**
   - add a card onto other food chains or food webs that have already been played.
6. Take turns until one student has played all of the cards in his or her hand.
7. Students should play all cards directly in front of them, even when adding onto someone else's food chain or food web. This will help scoring at the end.
8. Points are counted when a player discards or plays the last card in his or her hand.
9. Students subtract points for any cards they are holding.
10. The winner is the student with the most points using the table below:

Table 1: Food Web Builder scoring

Card	Points
Producer	1
Primary consumer	2
Secondary consumer	3
Top consumer	4
Sun	5
Decomposer	5

4. Using your model

a. Write one food chain sequence played in the game. Name each species in the chain and use arrows between the names to show the flow of energy up the chain.

b. Explain the importance of the Sun and photosynthesis to the flow of energy in all food chains.

c. Decomposers are often left out of food chains and food webs. Why are decomposers important in every ecosystem?

d. Divide up the cards so that each student has all the cards for one ecosystem. Use the cards to complete a food web. Remember a food web is more than one food chain linked together. Be sure to include names and energy niches, and draw arrows.

5. Constructing explanations

Draw and complete an energy pyramid with four levels using your game cards. Answer the following questions about how the game models real ecosystems.

a. What pattern do you see in all four ecosystems card sets that can help you predict interactions among species in other ecosystems, such as a pond or rainforest? Explain.

b. Which energy niche has the fewest cards? Which niche has the most? How does the number of cards for each niche and the shape of the pyramid model the actual transfer of matter and energy in ecosystems?

c. How are the point values for a producer versus a top consumer shown in the pyramid?

5B Testing Pollutants

How will similar populations react to different pollutants?

Human activities affect ecosystems in both positive and negative ways. One negative effect is pollution. A <u>pollutant</u> is a variable that causes harm to an organism. Pollutants enter ecosystems naturally and through human activities. For example, volcanic eruptions are a natural source of sulfur dioxide. Coal-burning power plants are a human source of this pollutant. In this investigation, you will test the effect of different levels of several pollutants on identical brine shrimp populations.

Materials
- 4 Brine shrimp populations in cups
- Isopropyl alcohol solution (50%)
- Milk solution (1%)
- Hydrogen peroxide solution (3%)
- Sugar solution (1%)
- Yeast solution
- Acetic acid (vinegar) solution (5%)
- Dropping pipettes
- Microscope, hand lens, digital camera
- Depression slides, cover slips
- Plastic tray

Safety Note: Wear goggles and an apron during this investigation.

WARNING — This lab contains chemicals that may be harmful if misused. Read cautions on individual containers carefully. Not to be used by children except under adult supervision.

1. Sampling populations

Start with four identical containers of brine shrimp. Each container should have the approximately the same concentration of brine shrimp and salt. By using four identical containers with about the same concentrations, you will have four similar populations that can be tested for their tolerance to a particular pollutant. Each group will test the effect of one pollutant and try to find an acceptable level (if any) that will not seriously harm a brine shrimp population.

1. Once your group has chosen a pollutant, decide how many drops will be added to each cup. Check with your teacher once you have agreed on an amount.
2. Label one of your containers "Cup 1 - Control Group." You will not add any pollutants to your control group.
3. Add the first amount of pollutant to another cup and label it "Cup 2." Be sure to identify the pollutant on the label and how much was added to the population.
4. Repeat step 3 for your other two cups.

Cup 1 CONTROL GROUP | Cup 2 20 drops | Cup 3 40 drops | Cup 4 60 drops

2. Thinking like a scientist

a. What is the experimental variable in this experiment?

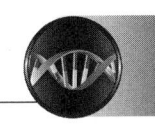

b. What are the control variables?

c. Will all the shrimp in each cup survive? Make a prediction for each cup.

d. Why is it important to add no pollutant to the control group?

e. How long do you think it will take for the first effects to be observable?

f. Why was it important to use the same sized containers with the same number of brine shrimp in each population?

3. Observing the populations

1. Confirm whether there are any shrimp alive using a hand lens or digital camera. Use technology to record the condition of the shrimp population by taking images or videos.
2. Using a dropping pipette, immediately capture a few brine shrimp from cup 1 and place on a depression slide. Examine them under a microscope. Keep sampling until you are confident of your results. Share your results with others in your group. If ANY shrimp are alive, record this in Table 1.
3. Repeat sampling in the other three cups.
4. Place your cups on a plastic tray marked with your group name. Store them away from direct sunlight.
5. Examine each cup for survivors for three more days and record your results.

Table 1: Brine shrimp population observations

Cup	Effect on populations			
	Day 1	Day 2	Day 3	Day 4
1				
2				
3				
4				

4. Analyzing data

a. Were there any survivors on day 4? If so, in which cups were there still living shrimp?

b. Do the cups all look the same or are there differences?

c. Which cup was most affected by the pollutant?

d. Which cup was least affected?

e. Does there seem to be an amount of the pollutant that does not affect the population?

5. Constructing explanations

a. What do the results of this investigation tell you about the pollutant you tested?

b. Did you observe any physical effects or changes when you observed your brine shrimp closely?

c. How could you modify the experiment to get a more precise amount for your acceptable pollutant level?

d. Do you think the pollutant you tested has the same effect on every other organism on Earth? Why or why not?

e. How can acceptable levels of a pollutant be determined for a particular environment?

f. How do you think scientists determine acceptable levels of pollutants for humans?

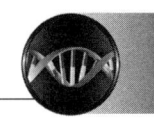

6A Environments Around School

What causes different microclimates within an ecosystem?

An amazing variety of physical conditions, called abiotic factors, can be found across Earth's biomes. These variables include precipitation, temperature, sunlight, wind, and soil. In this investigation, you will learn that even within one biome, physical conditions vary enough to create *microclimates*—conditions within one ecosystem that have unique physical characteristics. These characteristics encourage the growth of certain types of plants and discourage others. As a result, microclimates are of great interest to gardeners. You can find examples of microclimates in your own schoolyard!

Materials
- Weather tools: thermometer, hygrometer, barometer, compass, anemometer
- Pencil
- Data sheet or science journal
- Trowel
- Metric ruler
- Magnifying glass/hand lens
- Plant identification guides or keys
- Digital camera

Safety Note:
- **Dress appropriately when working outside.**
- **Never place any plant part in your mouth.**
- **Never rub any sap or fruit juice into your skin or an open wound.**
- **Never eat food after handling plants without first scrubbing your hands.**

1. Planning your investigation

1. Gather your materials.
2. Your teacher will assign a specific schoolyard site for your group to visit.

2. Gathering data

Find a comfortable place to stand or sit. In silence, make your own observations of the site for five minutes. Be sure to record the site name, the time of day, and the date. Write down everything you observe with your eyes, ears, and nose. Try not to disturb the site. You do not need to collect any plant or soil samples during this investigation. Record your observations in a science journal or separate paper using the format below.

Site Name:

Time of Day and Date:

Observations:

1. Estimate the amount of sunlight hitting the ground in your area and record it in Table 1.
2. Use the weather tools to collect data on temperature, humidity, and barometric pressure. Follow teacher or package directions before operating each device.

Table 1: Microclimate data

Variable	Data
Time of day:	
Light conditions: (full sun, mostly sunny, mostly shady, very shady)	
Temperature:	
Humidity:	
Barometric pressure:	
Wind direction:	
Wind speed:	

3. The Beaufort wind scale is used by meteorologists to observe wind in large areas. Use the scale to record observations. Then use the anemometer to estimate wind speed in your microclimate.

Beaufort wind scale (courtesy of NOAA Storm Prediction Center)

Force	Wind speed (knots)	Classification	Appearance of wind effects on land
0	less than 1	Calm	Calm, smoke rises vertically
1	1-3	Light air	Smoke drift indicates wind direction, wind vanes are still
2	4-6	Light breeze	Wind felt on face, leaves rustle, vanes begin to move
3	7-10	Gentle breeze	Leaves and small twigs constantly moving, lightweight flags are extended
4	11-16	Moderate breeze	Dust, leaves, and loose paper lifted, small tree branches move
5	17-21	Fresh breeze	Leaves in trees begin to sway
6	22-27	Strong breeze	Larger tree branches moving
7	28-33	Near gale	Whole trees moving, resistance felt when walking against wind
8	34-40	Gale	Whole trees in motion, stiff resistance felt when walking against wind
9	41-47	Strong gale	Slight damage to structures occurs, slate blown off roofs

Investigation 6A   *Environments Around School*

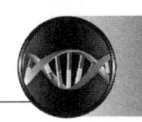

Beaufort wind scale (courtesy of NOAA Storm Prediction Center)

Force	Wind speed (knots)	Classification	Appearance of wind effects on land
10	48-55	Storm	Seldom experienced on land, trees broken or uprooted, considerable damage to structures.

4. Use a trowel and metric ruler to dig down about 20 centimeters into the soil. Use a magnifying lens to observe a small amount. Circle the soil type you observe in Table 2. When you return to class, your teacher may help you obtain more evidence for classifying it.

5. Make observations of the vegetation (plants) in your area. Circle the types you find in Table 2. Count or estimate the number of each type of plant and record in column 2. Use a magnifying lens to examine fine details on stems and leaves. If possible, take photos of each plant to review later. When you return to class, use a field guide provided by your teacher to identify one plant.

Table 2: Soil and vegetation data

Variable	Data
Soil Type: Sand Silt Clay Additional soil type:	
Vegetation: Trees Shrubs Flowering plants Grasses Ferns Mosses Aquatic plants: reeds, lilies, etc. Additional vegetation:	

Communicating your data

1. Record your group data on the class data table that has been set up by your teacher.
2. After all of the group data has been recorded, study the data, and answer the following questions.

4. Analyzing and interpreting your data

a. Which variable showed the greatest difference among the sites?

b. Which variable showed the least difference/was most similar among the sites?

c. Study the soil types and the vegetation of the different sites. Is there any relationship between the type of soil and the kinds of vegetation at the sites? Why or why not?

d. Study the light conditions and the vegetation of the different sites. Do you think there is any relationship between the amount of light an area receives and the vegetation in a given area? Why or why not?

e. Study the temperature and the light conditions of the different sites. Do you think there is any relationship between the amount of light a given area receives and the temperature? Why or why not?

f. Study the humidity and the light exposure of the different sites. Do you think there is any relationship between the humidity and the light in a given area? Why or why not?

5. Extending your investigation

a. Interview the person responsible for your school grounds. Ask him or her about the schoolyard microclimates and what they have observed, and how decisions are made about what to plant on the school grounds.

b. Invite a landscape architect to speak to your class about how microclimates play a role in landscape design.

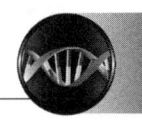

6B Field Study

How can a field study be useful?

Have you ever wondered how scientists study ecosystems and the animals that live there? An important way scientists gather this information is through careful observations. In this investigation, you will continue the previous study or begin one in a new community. This time you will focus on animal life. You will conduct a field study of that area, share your data with the entire class, and together, create a field guide to the schoolyard.

Safety Note:
- Dress appropriately when working outside.
- Never place any plant part in your mouth.
- Never rub any sap or fruit juice into your skin or an open wound.
- Never eat food after handling plants without first scrubbing your hands.

Materials
- Thermometer
- Hygrometer
- Metric tape and ruler
- Data sheet or science journal
- Clipboard
- Pencil
- Magnifying glass/hand lens
- Digital camera
- Binoculars (optional)
- Collecting jar (optional)
- Graph paper
- Animal field guides for a local area

1. Planning your investigation

1. Prepare to work outdoors with proper clothing and equipment.
2. Your teacher will assign a specific schoolyard site for your group to visit.

2. Making a prediction

a. How do you think a field study can be useful to an organization, such as a school?

b. What types of animals do you expect to find in your field study? Explain your predictions.

3. Gathering data

1. Find a comfortable place to stand or sit. In silence, make your own observations of the site for five minutes. Be sure to record **the site name, date, and time of day.** Write down everything you observe with your eyes, ears, and nose. Try not to disturb the site.

2. Record your initial observations in a science journal or separate paper using a similar format to what is shown in Investigation 6A.

3. Use the weather tools to collect data on sun exposure, temperature, and humidity. Follow teacher or package directions before operating each device. To save time, assign different tasks to members of your group.

4. Create a map of the area that you are to study. Draw its boundaries and include features such as large rocks, buildings, sidewalks, and so on. Try to draw your map to scale by using a relationship such as: *1 square on your graph paper = 1 square meter (or other unit)*

5. Measure the size of objects within your area using a cloth or retractable tape measure. Include your scale and label all features on your map.

6. You will conduct a field study to observe as many animals as you can find. During your time outside, spend some time walking and looking, as well as just standing still or sitting. Be sure to observe the ground and high above (such as treetops). If the ground is covered with leaves or gravel, be sure to examine it closely using a magnifying lens.

7. Make note of each animal that you observe in Table 1. If you do not know the name of the animal, include a description and a sketch or photo. Count the number of individuals present if there is more than one of these animals. If the number is too large to count, estimate by counting the number in a small area and multiplying. Identify the location(s) where you found the organism on your map using a letter or number code. Record any other important observations such as behavior. If you find more animals than there is space for in Table 1, create your own chart and continue.

Table 1: Field study data

Name of animal (Sketch if unknown)	Number of the same animal found	Map location	Other observations

4. Communicating your data

1. Share your individual results in your group and share group data with the rest of the class.
2. Use the field guides provided by your teacher to name unidentified organisms as you share.
3. Classify each organism found by your class into the groups shown in Table 2. Write the name of the organism and the number found in the appropriate columns of the table.

Table 2: Classifying animals

Type of animal	Animal's common name	Number of animals
Mammal		
Bird		
Reptile		
Bird		
Amphibian		
Arthropod		
Segmented worm		
Mollusk		

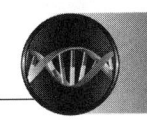

5 Arguing from evidence

a. Did you see the organisms that you predicted? Explain why or why not.

b. What factors might have influenced your field study? How might you make the results more inclusive of all of the organisms found in the schoolyard?

c. Did other groups report similar findings? Why or why not?

d. Create a bar chart to show the total number of organisms in each category from Table 2. Be sure to include all the important parts of a graph.

6 Extending your investigation

Create a class *Field Guide to the Schoolyard*. Each student will be assigned an organism from your class list to create a one-page field guide entry. Include the following information:

- Common name and scientific name of your animal
- Description (size, color, how it moves, and other characteristics)
- A sketch, image from the Internet, or picture from a digital camera
- Niche (role in the ecosystem)
- Nutrition (what it eats and how it gets its food)
- Habitat
- Range

7A Examining Onion Tissue

What evidence supports the idea that plant tissues have functions and different structures?

Plants have adapted different ways to store food and water for survival. Onion bulbs contain modified roots, stems, and leaves. One special tissue layer is easy to view under a microscope. In this investigation, you will peel a thin layer from a piece of onion. You will stain that layer of tissue on a microscope slide. Then you will examine your slide under a microscope and make sketches of what you see.

Materials
- Compound light or digital microscope (400x)
- Glass slides
- Cover slips
- Tweezers
- Piece of yellow or white onion
- 2% Potassium iodide stain in dropping bottle

Safety Note: Wear gloves, goggles, and an apron while preparing the slides.

WARNING — This lab contains chemicals that may be harmful if misused. Read cautions on individual containers carefully. Not to be used by children except under adult supervision.

1. Preparing a wet-mount of plant tissue

Onions have many layers. The inner surface of each layer has a thin layer of tissue that's easy to peel off. Since it is almost transparent, you will need to apply a stain so you can see structures under a microscope. Follow the procedures below to make a slide of onion tissue.

1. Place a drop of iodine stain onto a slide. If you cannot see light through the stain, add one drop of water.
2. Using the tweezers, gently peel the thin layer of tissue off the inside of a small piece of onion.
3. Using the tweezers, gently lower the onion tissue onto the slide. Be careful not to fold the tissue.
4. Use the tweezers to place a cover slip over the onion tissue.

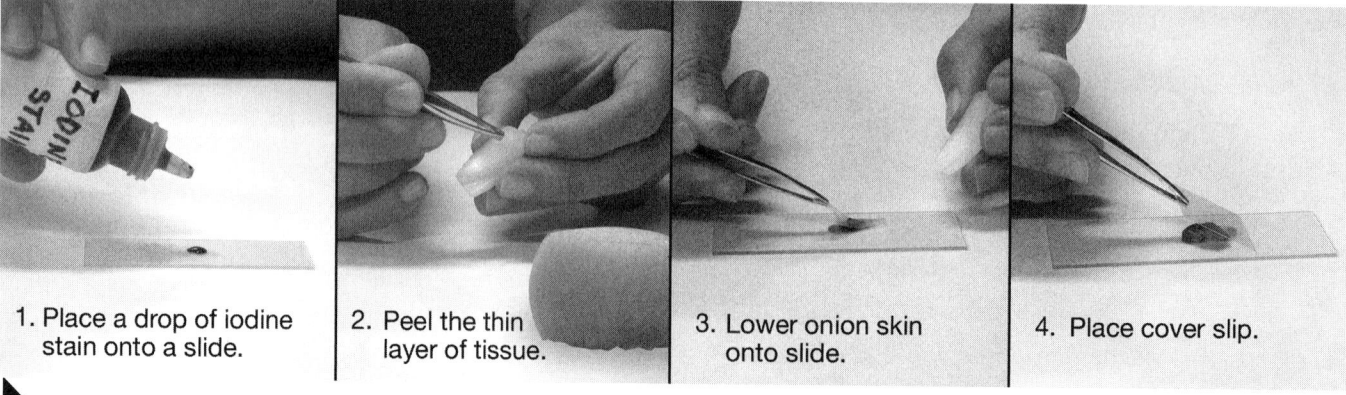

1. Place a drop of iodine stain onto a slide.
2. Peel the thin layer of tissue.
3. Lower onion skin onto slide.
4. Place cover slip.

2. Stop and think

a. What is the purpose of the iodine stain?
b. Recall from Chapter 2 how living things are organized. Of what are tissues made?
c. What is a tissue? What is the next level of organization above tissues? What is the level of organization below tissues?

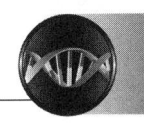

Investigation 7A   *Examining Onion Tissue*

3 Observing plant tissue under a microscope

1. Lower the stage on your microscope to its lowest point.
2. Switch to the low power objective lens (4x).
3. Place your slide on the stage and secure it with the clips.
4. Bring the slide into focus with the coarse focus (larger) knob. Center the tissue in the circular field of view.
5. Make a detailed sketch of what you see in the circle provided. Record your observations.
6. Lower the stage, switch to medium power, and repeat steps 4 and 5.
7. Lower the stage, switch to high power, and repeat steps 4 and 5.

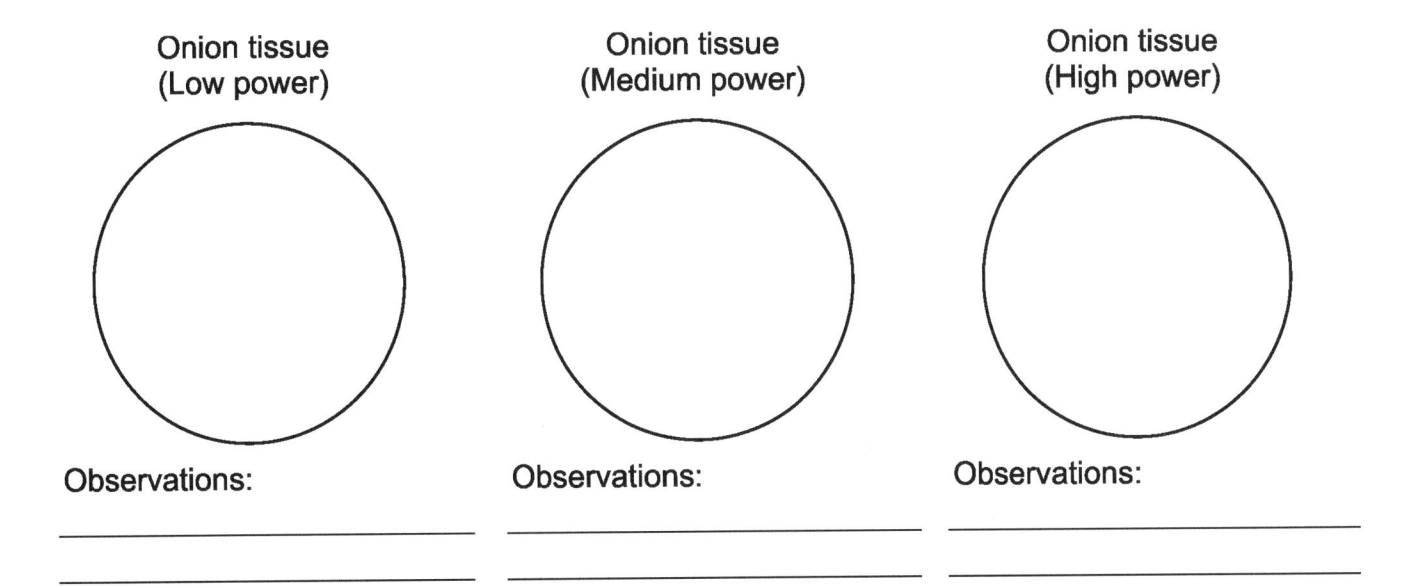

Onion tissue (Low power)

Onion tissue (Medium power)

Onion tissue (High power)

Observations: _____

Observations: _____

Observations: _____

4 Arguing from evidence

a. Based on your sketches and observations, what structures make up onion tissue?

b. The field of view stays the same as you change from low to medium or high power. How does increasing the magnification affect the amount of tissue you see?

c. How can you tell where one square structure ends and another begins? What do the individual structures you observed have in common?

d. When Robert Hooke looked at cork under a microscope in 1663, he called each of the square structures a *cell* because they reminded him of tiny rooms. Do your observations of onion cells agree with his? Explain why or why not.

e. Look at the diagram of a plant cell in Chapter 7 of your textbook. Which structures can you identify in your onion cells? Label them on your sketches.

7B Animal and Plant Cells

What are the differences between animal and plant cells?

In this investigation, you will compare animal cells (your own epithelial cells and prepared slides of muscle tissue cells) and plant cells (live Ulothrix—an algae that is closely related to plants—and the onion cells you observed in the last investigation).

Materials

- Safety goggles, apron, and gloves
- Compound light or digital microscope (400x)
- Glass slides
- Cover slips
- Tweezers
- Colored pencils
- Flat toothpicks
- Prepared slide of muscle tissue
- Live Ulothrix
- Methylene blue solution (0.3%)

Safety Note: Wear gloves, goggles, and an apron when preparing slides.

WARNING — This lab contains chemicals that may be harmful if misused. Read cautions on individual containers carefully. Not to be used by children except under adult supervision.

1. Observing animal cells

The cells that line the inside of your mouth are called *epithelial cells*. This tissue layer protects the mouth from constant abrasion. These cells are easy to collect and observe. Follow the procedures below.

1. Place a small drop of methylene blue stain onto a clean slide.
2. Get a clean flat toothpick. Wet the wide end of the toothpick using tap water.
3. Gently scrape the inside of your cheek with the wide end of the toothpick. DO NOT USE FORCE!
4. Place the toothpick into the stain on the slide and gently swirl to mix the cheek cells with the stain. Dispose of the toothpick as directed by your teacher. DO NOT REUSE THE TOOTHPICK.
5. Using tweezers, gently place a cover slip on top of the methylene blue solution as shown.
6. Place the slide on the microscope stage and observe under low power, medium power, and high power. Sketch what you see and record your observations in Table 1.
7. Dispose of the cheek cell slide as directed by your teacher.
8. Obtain a prepared slide of muscle tissue cells. Place the slide on the microscope stage and observe under low power, medium power, and high power. Sketch what you see and record your observations in Table 1.

1. Place a drop of stain on slide.
2. Wet wide end of a flat toothpick.
3. GENTLY scrub the inside of your cheek. *DO NOT SCRAPE WITH FORCE!*
4. Swirl toothpick in stain. Dispose of toothpick as directed.
5. Lower cover slip onto slide.

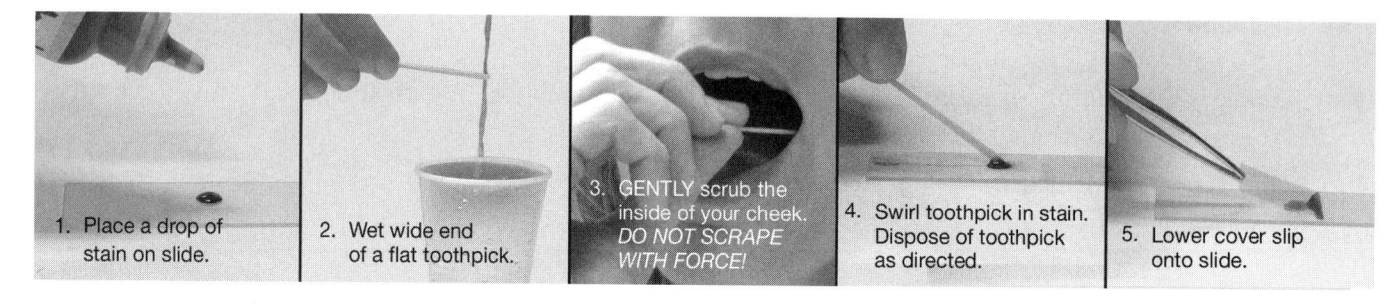

Investigation 7B *Animal and Plant Cells*

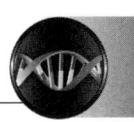

Table 1: Animal cell sketches and observations

Cheek cell (4X)	Cheek cell (10X)	Cheek cell (40X)
Observations:	Observations:	Observations:
Muscle tissue (4X)	**Muscle tissue (10X)**	**Muscle tissue (40X)**
Observations:	Observations:	Observations:

2. Constructing explanations

a. What is the purpose of adding methylene blue to the cheek cells?

b. How are the cheek and muscle cells different? How are they similar?

c. Look at the diagram of an animal cell in Chapter 7 of your textbook. Which organelle was most visible in the cheek cells you observed under the microscope?

d. How many individual cheek cells fit across the field of view on medium power (10x)?

e. Identify the nucleus, cell membrane, and cytoplasm in one of the cheek cells under high power. Label them on your sketch. Do these structures all have the same job in cheek cells?

f. How many individual muscle cells could you see under medium power (10x)?

g. Identify the nucleus and cell membrane in a muscle cell. Label them on your sketch. Compare the shape of a cheek cell to a muscle cell. Do these cells have the same function in humans?

41

3 Observing Ulothrix cells

1. Place a drop of water onto a clean slide.
2. Place some Ulothrix filaments into the drop of water.
3. Using tweezers, gently place a cover slip onto the filaments.
4. Examine the Ulothrix under low, medium, and high power.
5. Sketch what you see and record your observations in Table 2.

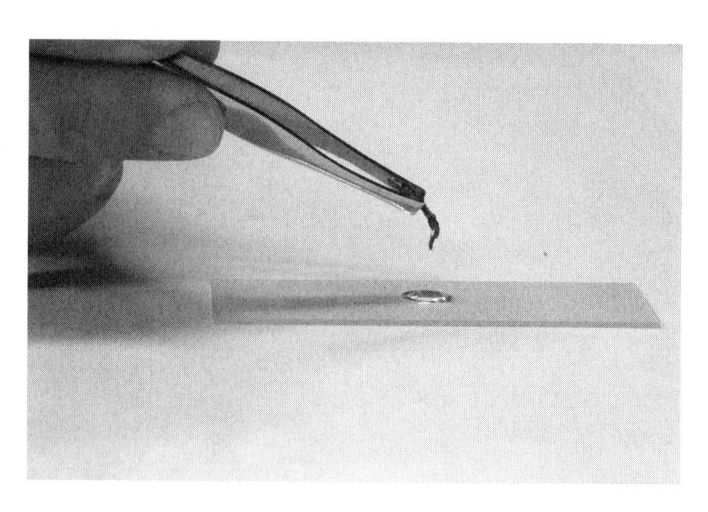

Table 2: Ulothrix cell sketches and observations

Ulothrix cells (4X)	Ulothrix cells (10X)	Ulothrix cells (40X)
Observations:	Observations:	Observations:

4 Constructing explanations

a. In Investigation 7A, you needed a stain to observe onion cells. Why didn't you need a stain to observe the Ulothrix cells?

b. Look at the diagram of a plant cell in Chapter 7, page 146 of your textbook. Which organelles can you identify in the Ulothrix cell slide?

c. How are the Ulothrix cells similar to the onion cells you observed in Investigation 7A? How are they different?

d. Label the following on your high-power sketch of Ulothrix cells: nucleus, cell wall, cytoplasm, vacuole, chloroplasts. What job do chloroplasts do in plant cells?

5 Communicating your results

a. Based on your sketches and observations, what structures do animal and plant cells have in common?

b. Based on your sketches and observations, what structures are found only in plant cells?

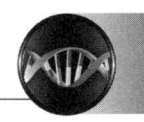

8A Diffusion and Osmosis

How does water move in and out of cells?

Water passes into and out of a cell by osmosis. *Osmosis* is the diffusion of water across a membrane from an area of higher water concentration to an area of lower water concentration. In this investigation, you will use an egg membrane as a model system and observe the effects of water movement when the egg is placed in different solutions.

Materials

- 2 500-mL Beakers or cups
- Marker or pen
- Masking tape
- Triple beam balance or digital scale
- 500 mL Vinegar (5%) in beaker or cup
- Plastic spoon
- 2 Small, fresh eggs
- Extra beakers/cups
- 400 mL Distilled water
- 250 mL Corn syrup
- Digital camera
- Plastic wrap

Safety Note: Wear gloves, goggles, and an apron during this investigation.

WARNING — This lab contains chemicals that may be harmful if misused. Read cautions on individual containers carefully. Not to be used by children except under adult supervision.

1. Planning the experiment

An egg can serve as a great model for demonstrating water movement. A fragile membrane just under the egg shell performs all the same functions as a cell membrane. To expose the membrane, we just need to remove the shell!

1. Obtain two 500-mL beakers. Number them "#1" and "#2." Write your initials or group number on each beaker.
2. Measure the mass of each beaker and record it in Table 1.
3. Obtain two eggs. Measure the mass of egg #1, record the mass, and place it into beaker #1.
4. Measure the mass of egg #2, record the mass, and place it in Beaker #2.
5. Fill each beaker with 250 mL of vinegar. Measure the mass of the contents of both beakers and record it in Table 1. How can you calculate the mass of the vinegar (the last column of Table 1)?

6. Allow the beakers to sit undisturbed overnight.

Table 1: Cell model mass (g) on day 1

Number	Beaker mass (g)	Egg mass (g)	Total mass (g)	Calculated vinegar mass (g)
Egg #1				
Egg #2				

7. On day 2, observe both beakers. Take pictures to include in your final report. Measure the total mass of each beaker again and record the new data in the first column of Table 2.
8. If any shell remains on the eggs, you will need to wait another day before proceeding.
9. If all shell is removed, drain the vinegar slowly and carefully from each beaker into another container or sink. Drain as much vinegar as you can, but **don't** break the membrane or accidentally dump the egg while working. A plastic spoon can help hold the egg in the beaker.

Table 2: Cell model mass (g) on day 2

Number	Total mass (g)	Beaker and egg mass (g)	Calculated vinegar mass (g)	Calculated egg mass (g)
Egg #1				
Egg #2				

10. Measure the mass of the beaker and egg again. Record it in Table 2. How much vinegar and shell did you remove? How can you calculate the new mass of the egg? (Hint: Can you use any data from day 1?) Complete Table 2.
11. Carefully rinse each egg with about 75 mL of distilled water. Drain as much water as you can. If you are successful, you will have two unbroken cell membranes ready for experimenting!

2. Developing a model

a. What was the effect of placing both eggs into vinegar overnight?

b. What happened to the mass of each egg after being placed into vinegar overnight?

c. If you think of the egg as a model of a "cell," what does the fluid inside the egg represent? What might the egg yolk represent?

d. In the next part of the investigation, you will place egg #1 into a beaker containing distilled water and egg #2 into a beaker containing corn syrup. Use what you know about osmosis to make a prediction about what you think will happen to the **mass** of each egg placed in distilled water or corn syrup overnight.

3. Planning and conducting the experiment

1. Pour 250 mL of distilled water into the beaker for egg #1 until the egg is completely covered.
2. Pour 250 mL of corn syrup into the beaker for egg #2 until the egg is completely covered. *If the egg floats, place a cup with water on top of the egg to keep it submerged.* Cover each beaker with plastic wrap and place them in a safe place overnight.

Investigation 8A   Diffusion and Osmosis

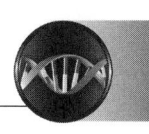

3. After 24 hours, slowly pour the water and syrup out of each beaker. Be very careful not to rupture the cell membrane or drop the eggs!
4. Observe each egg and record your final measurements in Table 3. If desired, take pictures.

Table 3: Cell model mass (g) on day 3

Number	Total mass (g)	Beaker and egg mass (g)	Calculated liquid mass (g)	Calculated egg mass (g)
Egg #1				
Egg #2				

4. Arguing from evidence

a. In question d of Part 2 above, you were asked to predict what would happen to each egg in the experiment. How did your results compare with your predictions?

b. Which beaker contained a greater concentration of water compared with the concentration of water in the egg: the beaker for egg #1 or the beaker for egg #2?

c. After 24 hours, did egg #1 contain more, less, or the same amount of water as it did before the experiment? What is your evidence?

d. After 24 hours, did egg #2 contain more, less, or the same amount of water as it did before the experiment? What is your evidence?

e. Use the terms *concentration*, *osmosis*, and *diffusion* to explain why water moved *into* one egg and *out of* the other.

5. Planning and conducting your own experiments

a. What would happen if you left the eggs in water and syrup for a longer period of time (another day, for example)? Is it possible to return the eggs to their original state before placing them into distilled water or corn syrup? Write a procedure and test your predictions.

b. What would happen if you placed the egg originally in corn syrup (egg #2) into distilled water? Could you restore the egg to its original state? Try this and see.

c. What would happen if you placed the egg originally in water (egg #1) into corn syrup? Would it shrink? Make a prediction and then try it.

8B Photosynthesis and Color

Does the color of light affect photosynthesis?

Living organisms, both plant and animal, contain chemicals known as *pigments*. A pigment's color is determined by the wavelengths of light that the pigment reflects. Plant leaves contain *chlorophyll*, a pigment that is vital to photosynthesis. In this investigation, we will find out which colors of light are needed by chlorophyll to sustain photosynthesis.

Materials

- 4 Small potted plants such as tomato or bean
- Plant grow light (75 W)
- Red light (75 W)
- Blue light (75 W)
- Green light (75 W)
- Four light fixtures
- Water
- Thermometer
- Cardboard dividers

1. Setting up

Plants use sunlight in their natural habitat to produce energy through the process of photosynthesis. Sunlight is a pure white light made up of all the colors together. What do you think would happen to plants if we didn't use white light but instead used individual colors of light?

1. Find a place in your classroom where you can set up the four lights and four small potted plants. You may need to place small cardboard dividers between the plants to make sure only the specific color of light you want falls on each plant.

2. Label each plant with the color of light. All of your plants should be in similar condition and approximately the same size. Record the height of each plant and the number of leaves on each stem.

3. Once the plant/light setup is in place, put a thermometer in one of the plant areas to monitor temperature. You won't keep track of the temperature, but you will check it periodically to make sure the plants don't get too hot. This may harm them and spoil the experiment.

Investigation 8B   Photosynthesis and Color

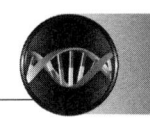

2. Stop and think

a. What is the experimental variable in this experiment?

b. What are the control variables?

c. Make a hypothesis stating how you think the color of light used will affect each plant. Think about the color of your plant. What color or colors is it reflecting? What color or colors is it absorbing?

3. Doing the experiment

1. The experiment begins when the lights are turned on. Discuss with your teacher if you will be using a 12-hour timer to turn the lights on and off, if this will be done manually, or if they will be on 24 hours a day.
2. Decide with your group if and/or how you want to water your plants.
3. Check on your plants each day and record your observations in your journal. Use a data table like Table 1 below to record your observations. Include a column to describe the initial condition of your plants.

Table 1: Plant growth/health data

	Day ___	
Color	Initial condition	Change in condition
Grow light		
Red		
Green		
Blue		

4. Thinking about what you observed

a. What color is your plant? If that particular color is getting to your eyes from the plant, is the plant reflecting or absorbing that color?

b. What is the plant doing with the colors it absorbs?

c. Which color(s) of light you tested seem to support photosynthesis?

d. Which color(s) of light you tested did not seem to support photosynthesis?

e. Starting from the sun, describe the process that allows you to see the color of your plant.

f. Why do certain colors of light support photosynthesis while others do not?

47

9A Protozoans

What are the characteristics of protozoans?

Most organisms in Kingdom Protista consist of a single cell. How do these organisms move and gather food? In this investigation, you will examine protozoans called amoeba, paramecium, and euglena using the light microscope. You will observe their structures and characteristics. You will also compare and contrast movement and feeding methods.

Materials

- Live euglena culture
- Live amoeba culture
- Live paramecium culture
- Glass slides
- Cover slips
- Corn syrup or methyl cellulose solution (1.5%)
- Digital or compound light microscope (400x)
- Toothpicks
- Dropping pipette

1. Preparing slides of live protozoans

Observing amoeba

1. Place a pipette into the bottom of the amoeba culture and gently depress the bulb. Extract a tiny amount of liquid and place it onto a glass slide.
2. Carefully place a cover slip on the slide, and place the slide under the microscope.
3. Using low power (4x), scan the slide for signs of slow-moving and transparent amoeba. Reduce light if needed.
4. Once you have found the amoeba, switch to the medium-power (10x) objective and focus again. Center the amoeba before changing to high power.
5. Make a sketch of what you observe under the microscope in Table 1.
6. Observe the movement of the amoeba and record your observations in Table 1.

Observing paramecium and euglena

1. Place a single drop of corn syrup or methyl cellulose solution onto a glass slide. Protozoans are very fast; corn syrup will slow the organisms so they may be viewed more easily under the microscope.
2. Place a single drop of paramecium culture onto the syrup on the slide.
3. Using a toothpick, carefully mix the water, corn syrup, and paramecium culture.
4. Carefully place a cover slip on top and place the slide under the microscope.
5. Using medium power (10x), bring the moving paramecia into view.
6. Once you have found a slow-moving paramecium, switch to the high-power objective (40x). You may have to adjust the diaphragm to control the lighting.
7. Make a sketch of what you observe under the microscope in Table 1.
8. Observe the movement of the paramecium and record your observations in Table 1.
9. Repeat the procedures using the euglena culture.

Investigation 9A *Protozoans*

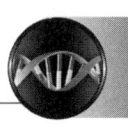

Table 1: Protozoa observations

Amoeba	Paramecium	Euglena
Observations:	Observations	Observations:

2. Constructing explanations

a. Describe the way each organism (amoeba, paramecium, and euglena) moves. Which ones have structures for movement? What do those structures look like?

b. Do any of the organisms have a structure for feeding? How do you think they obtain food?

c. Use the diagrams (right) to help you identify the structures of each organism. Label the structures you observed in your sketches.

3. Obtaining and evaluating information

Protozoans are everywhere and will make their homes in a variety of water sources. Some protozoans cause sickness in humans. One species, called *Giardia lamblia,* is closely related to paramecium. It lives in freshwater sources. Research *Giardia* and write a short report. Use the key word "Giardia" in your Internet search. Cite all sources and include the following information in your report:

a. Where is *Giardia* found? How do people contract the disease?

b. What are the symptoms of the disease caused by *Giardia*?

c. How is the disease treated?

d. How can the disease caused by *Giardia* be prevented?

9B Investigating Pond Water

Which microscopic organisms are found in pond water?

If you took a single drop of water from a pond and looked at it under a microscope, you'd be lucky to find anything. To increase the number of organisms in your sample you can create a hay infusion. A hay infusion is a culture that uses water collected from a pond, stream, creek, or puddle and dried grass. You simply place dry grass in a container of water and allow the culture to sit for approximately one week. At the end of the week, the culture will be teeming with different types of microscopic organisms. In this investigation, you will make a hay infusion and then, using what you have already learned, try to identify the microorganisms.

Materials

- Glass jar
- Pond or creek water
- Hay or dry grass
- Milk
- Yeast
- Glass slides
- Cover slips
- Corn syrup or methyl cellulose solution
- Digital or compound light microscope
- Dropping pipette
- Pond organisms chart
- Plastic wrap

1. Setting up

Adding milk | Adding yeast

1. Place pond water into the glass jar until the jar is about half filled.
2. Add several pieces of cut hay or dry grass to the jar.
3. Add a teaspoon of milk and a grain of yeast.
4. Loosely cover the jar with plastic wrap.
5. Place the jar in a warm place, in front of a window or in an incubator.
6. Check the hay infusion periodically. Although it will take about a week for the organism growth to peak, spot checking the culture will ensure you see lots of different organisms.

Safety Note: The hay infusion will contain a large number of bacteria. Make sure you wash your hands completely after handling the hay infusion, samples, and slides.

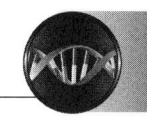

2. Making a prediction

a. What types of organisms do you think you will find under the microscope after a week?

b. Why is it important not to tighten the lid of the jar completely?

3. Making observations

1. Remove the jar's lid. Take a small drop of the water from the top of the jar and place it onto a clean glass slide.
2. Add a drop of corn syrup or methyl cellulose solution.
3. Carefully place a cover slip on top, and place the slide under the microscope.
4. Using the low-power objective, bring the sample into focus. Change to medium power (10x) and move the slide to a region of slow-moving organisms. Focus.
5. Once you have found the correct plane of focus, switch to the high-power objective. You may need to do this several times to find a single organism to observe.
6. Make a sketch of what you observe under the microscope. It is possible you will see a number of organisms, and this is good. Make a detailed sketch of every organism you see.
7. Once you have completed steps 1-7, wash and dry the slide.
8. Repeat the procedures with drops of water taken from different areas of the jar.
9. Once you have completed your drawings, try to identify the organisms you have observed in your hay infusion. A simple pond organism chart will be provided by your teacher. Use other resources such as books or the Internet, if needed.

Table 1: Hay infusion observations

Sample location (top, middle, bottom)	Sketch of the organism	Name of organism

4. Communicating your results

a. How many different types of organisms did you identify?

b. What type of information did you use to help you identify the different organisms in your hay infusion?

c. Was there a difference between the types of organisms you observed at the bottom, middle, and top of the jar?

d. Which location contained the largest number of different organisms?

e. Which organisms belong to the Kingdom Protista? Which organisms do not?

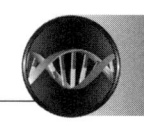

Investigation 10A   10A Observing the Cell Cycle

10A Observing the Cell Cycle

How much time is spent in different stages of the cell cycle?

The life cycle of a cell is called the *cell cycle*. Interphase is the period a cell spends growing and performing its functions. *Mitosis* is the part of the cell cycle where the chromosomes are divided into two new nuclei. Two daughter cells are formed, each containing a complete set of chromosomes. Mitosis has 4 phases: prophase, metaphase, anaphase, and telophase. After mitosis, the cytoplasm and its contents divide, separating the daughter cells (cytokinesis). In this investigation, you will determine the percentage of time a cell spends in interphase and the four stages of mitosis.

Materials

- Prepared onion root tip slide (*Allium*)
- Prepared roundworm slide (*Ascaris*)
- Textbook
- Pencil for sketching
- Graph paper
- Digital or compound light microscope (400x)

1 Planning your investigation

1. Set up a microscope and turn on the light.
2. On the stage, place a slide containing a stained preparation of *Allium* (onion root tip).
3. Locate the growth zone of the onion root tip. It is just above the root cap at the very end.
4. Focus on low power, and then switch to medium or high power. Below are pictures of the four stages of mitosis. Use them to help you identify the different stages on the microscope slide.

Phases of mitosis (prepared slides of *Allium*)

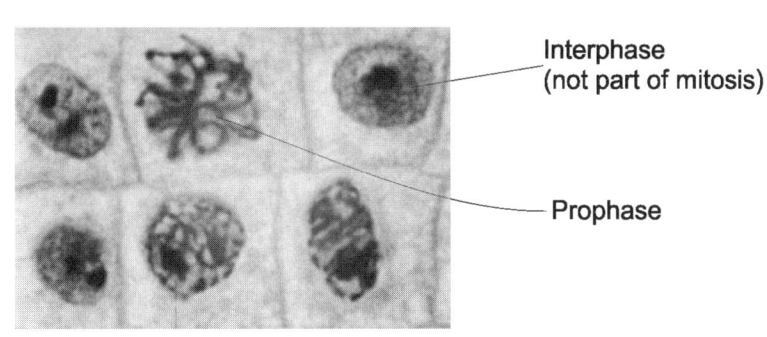

Interphase (not part of mitosis)

Prophase

Interphase and Prophase (40X)

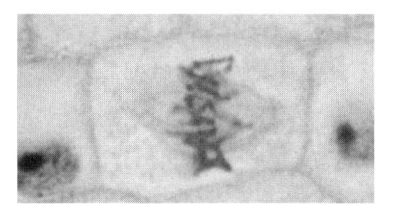

Metaphase (40X)

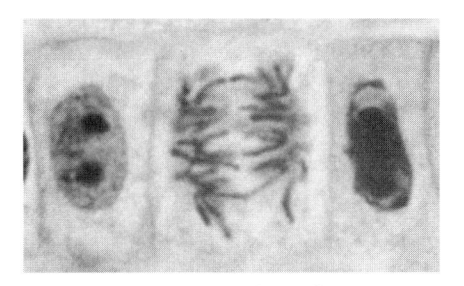

Anaphase (40X)

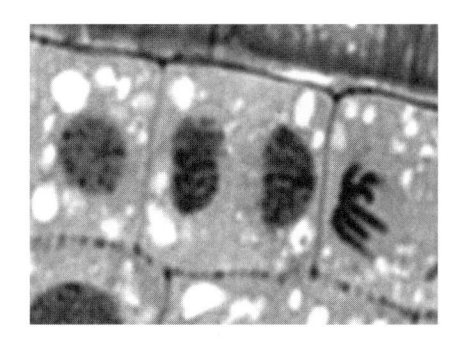

Telophase (40X)

53

2. Making a prediction

a. What percentage of the cells do you think will show each stage of the cell cycle (interphase, prophase, metaphase, anaphase, and telophase)?

b. Why is it important to recognize the differences in the appearance of the chromosomes at different stages?

3. Making observations

1. Count the number of cells in interphase, prophase, metaphase, anaphase, and telophase. You'll have to move the slide around several times until you have covered the entire growth region. Since cells on the slide were all preserved at the same time, comparing large numbers of cells at each stage can allow you to infer the amount of time spent there. Record your data in Table 1 below.

2. Add up the total number of cells you counted and enter that value in Table 1.

3. To calculate the percentage of time in the cell cycle, use this formula:

(Number of cells in phase ÷ total number of cells) × 100

Record your values in Table 1.

Table 1: Number of cells and percentages

Stage of cell cycle	Number of cells in stage	Time spent in stage (%)
Interphase		
Prophase		
Metaphase		
Anaphase		
Telophase		

4. Constructing and using models

a. Make a line or bar graph of the data in Table 1. Place time spent on the x-axis and number of cells on they y-axis.

b. In which stage do the highest percentage of cells spend their time? Of the four phases of mitosis (excluding interphase), which phase takes the most time to complete?

c. Cytokinesis is the stage in the cell cycle that happens right after mitosis. In this stage, the cytoplasm is divided and the two daughter cells are separated. See if you can find an example of cytokinesis on your *Allium* slide.

d. Obtain a prepared slide of animal roundworm cells (*Ascaris*). Try to identify and sketch cells in different stages of the cell cycle. Make a sketch of a cell in interphase, prophase, metaphase, anaphase, and telophase. Try to identify and sketch cells in cytokinesis.

e. How is cytokinesis different in the animal cell?

Investigation 10B Modeling Mitosis and Meiosis

10B Modeling Mitosis and Meiosis

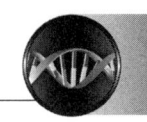

How do sex cells end up with a haploid set of chromosomes?

You may have seen fruit flies buzzing around a bowl of fruit. They are tiny, but if you look closely, you may see their red or white eyes. Like all living organisms, fruit flies grow and reproduce. The diploid number of chromosomes in fruit flies is 8. In this investigation, you will simulate mitosis and meiosis in fruit flies. Through modeling you will learn how fruit fly sex cells end up with only 4 chromosomes.

Materials

- Pipe cleaners of 2 different colors cut to 4 different lengths (see photo below)
- Poster board or large piece of paper
- Marker
- O-shaped cereal
- Textbook

1. Creating your model

1. Copy the chart (right) onto a piece of poster board using a marker. The circles represent a fruit fly body cell in different stages of the cell cycle and mitosis.

2. Your teacher will give you a set of pipe cleaners to represent chromosomes. One color will represent chromosomes contributed by the fly's mother and the other color represents the chromosomes contributed by the fly's father. Since chromosomes occur in homologous (one from each parent) pairs, use the same length of pipe cleaner for each homologous pair. You should create two sets of four different lengths of pipe cleaners.

3. Begin by assembling a *diploid* set of chromosomes for a fruit fly as they exist during most of interphase. A diploid set contains pairs of homologous chromosomes. Each chromosome contains unique information in its DNA. You will have an extra set of each length and color left over. Here is a diploid set:

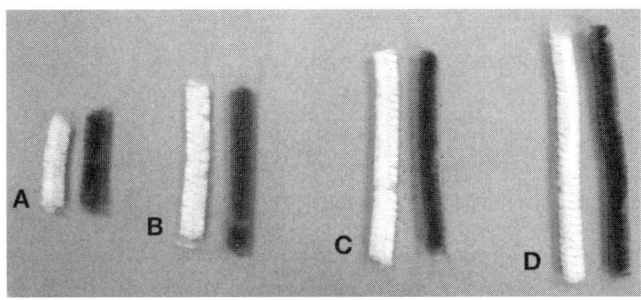

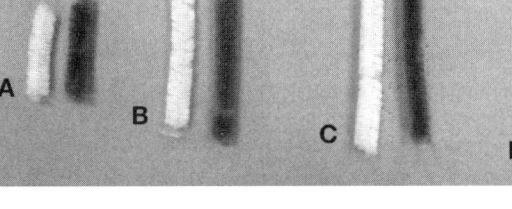

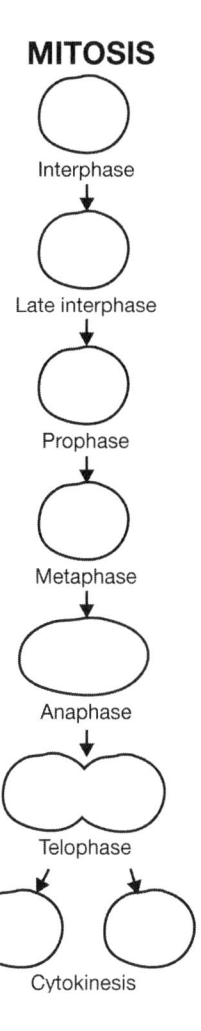

MITOSIS

Interphase

Late interphase

Prophase

Metaphase

Anaphase

Telophase

Cytokinesis

2. Using your model

a. What is the diploid (total) number of chromosomes in a fruit fly body cell?

b. How many homologous pairs of chromosomes does a fruit fly have?

c. In the chart (right), name the steps that are part of mitosis.

d. Which steps are part of the rest of the cell cycle?

55

3. Modeling mitosis

1. In late interphase (the second circle on your poster), the amount of DNA doubles. That means each chromosome now doubles. Select a matching pipe cleaner (same length and color) for each chromosome and slide both through a cereal ring. You now should have a set of eight doubled chromosomes arranged in homologous pairs.

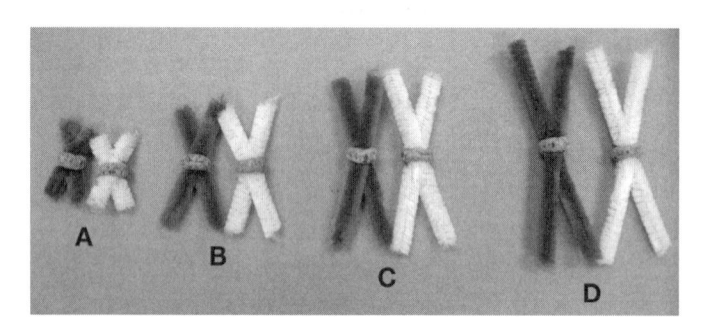

2. Review the mitosis diagram in Chapter 10 of your textbook, and move the chromosomes through the rest of the steps on your poster.

4. Evaluating information

a. Fill in Table 1 with the correct information.

Table 1: Cell cycle and mitosis in fruit flies

Step	Number of cells	Number of chromosomes in each nucleus	Number of homologous pairs in each nucleus
Interphase			
Cytokinesis			

b. What is the purpose of mitosis?

c. A diploid set of human chromosomes contains 23 homologous pairs (46 chromosomes). Fill in Table 2 with the correction information regarding human body cells.

Table 2: Cell cycle and mitosis in humans

Step	Number of cells	Number of chromosomes in each nucleus	Number of homologous pairs in each nucleus
Interphase			
Cytokinesis			

d. Why is it necessary to double the amount of genetic material before mitosis begins?

e. The two daughter cells end up with an exact copy of the genetic material from the parent cell. How does your simulation support this statement?

Investigation 10B Modeling Mitosis and Meiosis

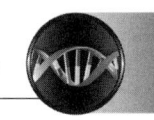

5. Modeling meiosis

<u>Meiosis</u> is the process of producing sex cells with a haploid set of chromosomes. <u>Haploid</u> means single—in this case, a single set of chromosomes. A haploid cell contains half the number of chromosomes as the original cell, one chromosome from each homologous pair.

1. Turn over your poster board and copy the chart (right) onto the other side. Fill the entire space with your drawing.
2. Begin by assembling a diploid set of chromosomes for the fruit fly as you did in Part 1. Place the chromosomes, in homologous pairs, in the first nucleus on your chart.
3. Like mitosis, before meiosis begins, the chromosomes double. Add a similar pipe cleaner to each chromosome and attach with a piece of cereal as you did in Part 1.
4. Unlike mitosis, which has only one cell division, meiosis has two divisions (meiosis I and meiosis II). Using the meiosis diagram in Chapter 10 of your textbook as a guide, move your chromosomes through meiosis I and meiosis II.
5. Fill in Table 3 below as you move your chromosomes through the chart.

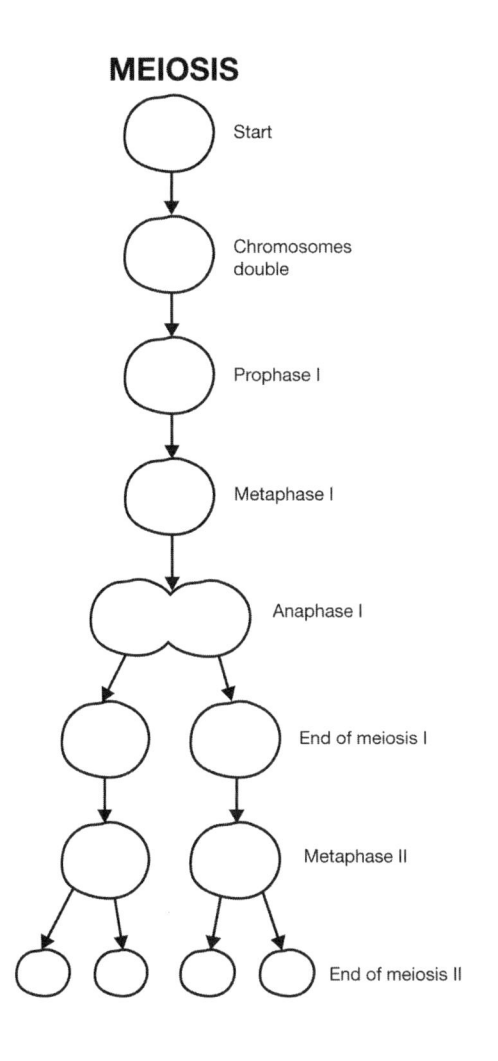

MEIOSIS

- Start
- Chromosomes double
- Prophase I
- Metaphase I
- Anaphase I
- End of meiosis I
- Metaphase II
- End of meiosis II

Table 3: Meiosis

Step	Number of cells	Number of chromosomes in each cell	Number of homologous pairs in each cell	Diploid or haploid number?
Start of meiosis				
End of meiosis I				
End of meiosis II				

6. Constructing explanations

a. What happens to the homologous pairs of chromosomes in meiosis I?

b. At the end of meiosis I, how does the number of chromosomes in each new cell compare to the original number of chromosomes in the parent cell? (Is it diploid or haploid?)

c. At the end of meiosis II, how does the number of chromosomes in each new cell compare to the original number of chromosomes in the parent cell?

57

7 Extending your investigation

If time permits, you can model what happens during fertilization.

1. Sketch the chart (right) onto a blank piece of paper.
2. Remove the chromosomes from a sex cell you created during the investigation, and combine this with the chromosomes of a sex cell from a different group. This is less complicated if you trade with a group that has different color pipe cleaners from yours. Place the correct number of chromosomes in each circle.

a. What is the end result of fertilization in terms of chromosome number?

b. Do the cells of the new organism have a diploid or haploid set of chromosomes?

c. What would happen if the sperm cell and egg cell did not reduce the number of chromosomes before fertilization? Model this.

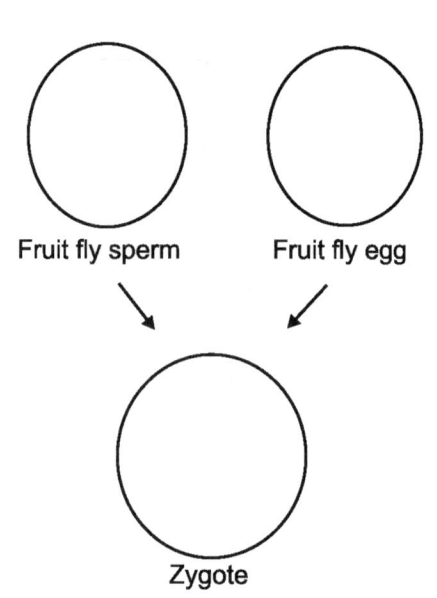

d. You have 23 pairs of chromosomes in your body cells. Why do your chromosomes occur in pairs? For each pair, where does each chromosome come from?

e. Recall that fruit fly body cells have four pairs of chromosomes and humans have 23 pairs. Some ferns contain 300 pairs of chromosomes. Does this mean fruit fly cells are less complicated than human cells, or fern cells are more complicated than human cells? Research number of chromosomes and complexity of organisms to learn more.

Investigation 11B *Crazy Traits*

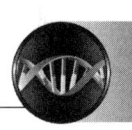

Table 1: Genotypes and phenotypes of offspring for Part 1

Trait	Allele from mother	Allele from father	Genotype	Phenotype
1. Gender				
2. Skin color				
3. Leg				
4. Foot				
5. Arms				
6. Hands				
7. Eye color				
8. Eyebrows				
9. Beak				
10. Ears				
11. Antenna				
12. Antenna shape				
13. Tail				
14. Wings				

3 Building your model

1. Once you have completed columns 2 through 4 of Table 1, use Table 2 (next page) to look up phenotype for each trait. Record the phenotype for each trait in column 5 of Table 1.
2. Once you have completed Table 1, your teacher will provide the parts to build your creature.
3. Carefully assemble your creature using tips provided by your teacher.
4. Give your creature a name and make it a name tag using a sticky note. Write the gender of your creature on the name tag.
5. Display your creature with models from other groups.

4 Arguing from evidence

a. Examine the creatures. Do any of them look exactly alike? Why or why not?

b. How does this investigation explain why siblings may resemble each other, but never look exact (unless they are identical twins)?

c. Count the number of males and number of females. Does the number of each match the chance getting a male or female predicted by a Punnett square? Why or why not?

d. Which trait(s) are examples of complete dominance? (Hint: Examine the genotype options for trait in Table 2.)

e. Which trait(s) are examples of incomplete dominance?

f. Which trait(s) are examples of codominance?

63

Table 2: Key to genotypes and phenotypes

Trait	Genotypes and phenotypes
1. Gender	*XX* - female *XY* - male
2. Skin color	*TT* - red *Tt* - purple *tt* - blue
3. Leg	*TT* - short *Tt* - short *tt* - long
4. Foot	*TT* - webbed *Tt* - webbed *tt* - talons
5. Arms	*TT* - long *Tt* - long *tt* - short
6. Hands	*TT* - paws *Tt* - paws *tt* - claws
7. Eye color	*TT* - red *Tt* - one red and one green *tt* - green
8. Eyebrows	*TT* - unibrow *Tt* - unibrow *tt* - separate
9. Beak	*TT* - trumpet *Tt* - trumpet *tt* - crusher
10. Ears	*TT* - elephant *Tt* - elephant *tt* - mouse
11. Antenna	*TT* - long *Tt* - long *tt* - short
12. Antenna shape	*TT* - knob *Tt* - knob *tt* - star
13. Tail	*TT* - long *Tt* - short *tt* - none
14. Wings	*TT* - no wings *Tt* - no wings *tt* - wings

Table 3: Offspring genotypes and phenotypes for Part 5

Trait	Genotype of mother for the trait	Genotype of father for the trait	Genotype of offspring (after flipping)	Phenotype of offspring
1. Gender				
2. Skin color				
3. Leg				
4. Foot				
5. Arms				
6. Hands				
7. Eye color				
8. Eyebrows				
9. Beak				
10. Ears				
11. Antenna				
12. Antenna shape				
13. Tail				
14. Wings				

Investigation 11B *Crazy Traits*

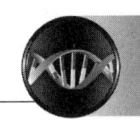

5. Extending your investigation

If time permits, work with another group whose creature is the opposite gender. Follow the steps below to create offspring of the couple.

1. Record the genotypes of each parent in the second and third column of Table 3 on the previous page.
2. First, toss for gender using the male and female sex chromosome coins.
3. For each trait, you'll need to use the correct egg and sperm coins for each parent. Use the data in Table 1 to find the parents' genotype for each trait. Then, select the egg and sperm coin that has the same alleles as the genotype. For example, if the father's genotype for skin color is *TT*, choose the blue sperm coin that has a capital *T* on both sides of the coin. If the mother's genotype for skin color is *tt*, choose the green egg coin that has a lower case *t* on both sides of the coin.
4. Place both coins in the cup, shake, and toss out onto the table. Record your results in the fourth column of Table 3.
5. Use Table 2 to look up the phenotypes. Record the phenotypes of the offspring in the last column of Table 3.

6. Constructing explanations

a. Which parent does your offspring share the most phenotypes with, the mother, father, or both equally?

b. Why did you need to choose different egg and sperm coins for each trait and for each parent?

c. What is the process taking place in cells that is modeled by flipping the black and red coins? (Hint: Recall Investigation 10B.) Why is this process important?

d. There is always a 50 percent chance of producing a male Crazy Creature. Argue why this statement is true. You may use a diagram to help support your argument.

e. In Part 1, you started off with both parents having identical genotypes for all traits. Use what you have learned in the investigation to argue why this is unlikely to occur in nature.

f. CHALLENGE! Make Punnett squares to show possible genotypes and phenotypes for each trait you flipped for in Part 5. For each, list the chances for each phenotype as a ratio and as a percent.

12A The DNA Molecule

What is the structure of the DNA molecule?

Deoxyribonucleic acid (DNA) is the hereditary molecule. DNA is made of individual units called <u>nucleotides</u>. Each nucleotide is made of a deoxyribose (sugar), a phosphate, and a base. The DNA molecule looks like a twisted ladder. The phosphate and sugar form the sides of the molecule. Each step contains a pair of bases held together by hydrogen bonds. There are four bases: thymine (**T**), adenine (**A**), guanine (**G**), and cytosine (**C**). **T** and **A** always pair up and **G** and **C** always pair up. In this investigation, you will model the structure of DNA using beads to represent the different parts of the DNA molecule.

Materials

- Red pop beads
- White pop beads
- Yellow pop beads
- Green pop beads
- Orange pop beads
- Blue pop beads
- Clear plastic connectors
- Cups to hold each bead color
- Colored pencils

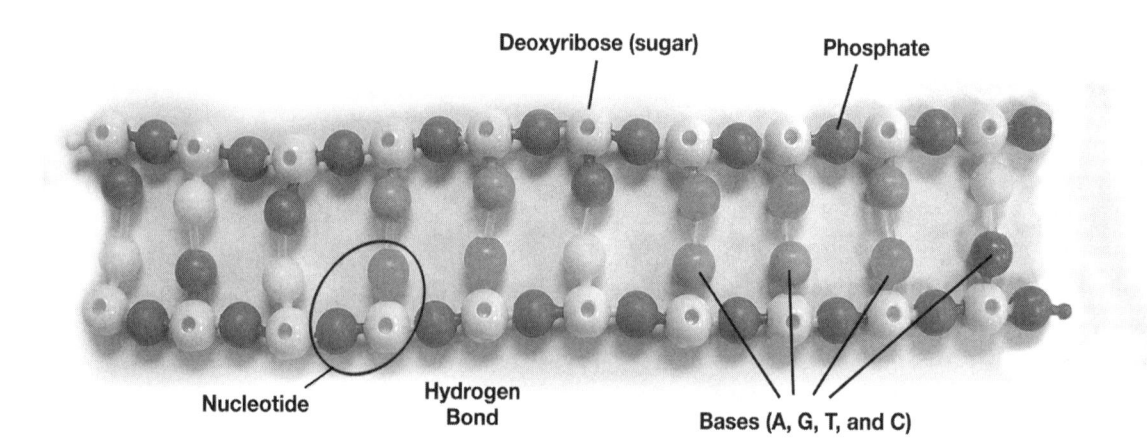

Setting Up

Gather the required materials. Use Table 1 as a key for creating your DNA model.

Table 1: DNA model components

Pop bead color	Molecule
Red	Phosphate
White	Deoxyribose (sugar)
	Bases:
Blue	Cytosine (**C**)
Orange	Guanine (**G**)
Yellow	Adenine (**A**)
Green	Thymine (**T**)

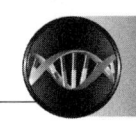

Investigation 12A The DNA Molecule

2. Developing a model

1. Begin by creating a nucleotide. Connect one red and one white bead and add any other color to make an L-shaped unit.
2. Instead of creating and connecting nucleotides, it is easier to build two separate strands to serve as DNA "backbones." Connect 10 **deoxyribose** (white) and 10 **phosphate** (red) beads in alternating order; do this twice to create two "backbones."
3. Lay the two strands so that they are parallel. One strand starts with a red bead and the other starts with red (see grahic on previous page).
4. Attach a **base** to each sugar on **one** of the phosphate-sugar backbones. We will call this the **original DNA strand**. At this point, the order of the bases does not matter. Use blue to represent **C**, orange to represent **G**, yellow to represent **A,** and green to represent **T** (Table 1).
5. Once you have created the original DNA strand, complete the first two columns of Table 2. Place the color of the bead followed by the letter of the base in order as they appear in your model.

Table 2: DNA molecule data

| Original strand || Complementary strand ||
pop bead color	base	complimentary base	pop bead color

6. Once you have recorded the sequence of bases on the original DNA strand in Table 2, determine the base sequence for the **complementary DNA strand**. Remember, **T** always pairs with **A,** and **G** always pairs with **C**.
7. In Table 2, record the name of the complementary base and the color of the corresponding pop bead.
8. Using the second phosphate-sugar backbone that you created in step 2, make the complementary DNA strand using your data from Table 2.
9. Once you have created the complementary DNA strand, use the clear plastic connectors to "bond" the base pairs together. These connectors represent the weak hydrogen bonds between complementary bases. During DNA replication, these bonds break to unzip the DNA molecule.
10. Using colored pencils, sketch and label your DNA molecule in Table 3. Circle and label one nucleotide.
11. Hold the model from the top, and gently twist the DNA ladder to the right. You should see that the DNA looks like a spiral staircase. The model now represents the **double helix** structure of DNA.

12. Your teacher may instruct the class to attach some or all of the DNA models together to make a longer strand of DNA. Do not do this unless instructed by your teacher.

Table 3: DNA molecule sketch

3. Using your model

a. Which molecules make up the backbone of the DNA molecule?

b. Why is DNA called "deoxyribonucleic acid"?

c. What type of bond keeps the bases paired together?

d. Which base always pairs with adenine? Which base always pairs with cytosine?

e. Compare your base sequence to others in your group. Are they identical? If so, what might an identical base sequence mean for a cell function?

4. Exploring on your own

Watson and Crick were awarded the Nobel Prize for their work in determining the structure of DNA. However, the work of many scientists led to the final determination of the structure of DNA. One such scientist was Rosalind Franklin. Do some research (library or Internet) to learn more about Rosalind Franklin. In a paragraph, discuss what she was studying and why this helped Watson and Crick determine the structure of DNA. Explain why her contribution might have gone unnoticed for so long.

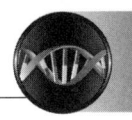

12B DNA Forensics

How can DNA be used to solve a "crime"?

You have learned that DNA is the molecule that carries hereditary information. Each individual is genetically unique, except in cases of identical twins, and as a result your DNA becomes your molecular fingerprint. This molecular fingerprint can be used to identify an individual with a great degree of certainty and is used in the field of forensic science. Scientists use DNA left at crime scenes to identify a person who may have been involved in a crime. In this investigation, you will use DNA to solve a fictional crime.

Materials
- Red pop beads
- White pop beads
- Yellow pop beads
- Green pop beads
- Orange pop beads
- Blue pop beads
- Clear plastic connectors
- Suspect DNA sequence on envelope containing evidence

1. Creating a model

Today, someone left your teacher an apple with a note, "You are a super teacher!" But between the time the apple was left and the time your teacher returned to class, a bite was taken out of it! Imagine that DNA has been collected from the "crime scene." A DNA sample has also been collected from suspects. Your lab group will help evaluate evidence by building a portion of a DNA molecule from one of the suspects. Then, you will compare your DNA model with others in the class. Can you solve the mystery?

1. Gather the required materials. Use Table 1 as a key for creating your DNA molecule.

Table 1: DNA model components

Pop bead color	Molecule
Red	Phosphate
White	Deoxyribose (sugar)
Blue	Cytosine
Orange	Guanine
Yellow	Adenine
Green	Thymine

2. Your teacher will give each group a sealed envelope. Do not open it until directed by the teacher. The DNA sample number is printed on the outside. Record your DNA sample number in Table 2.

Table 2: DNA sample data

DNA sample number:	
Number of bases:	
Number of deoxyribose (sugar) molecules:	
Number of phosphates:	

3. Using the base sequence on the outside of the envelope provided, count the total number of bases needed and indicate this number in Table 2.

4. Recall that the bases attach to the deoxyribose (sugar) molecules. Calculate the number of sugar molecules that will be needed and record this in Table 2. Since the backbone is alternating phosphate and sugar molecules, calculate the number of phosphates that will be needed. Record this in Table 2.

5. Using Table 2, create two phosphate-sugar backbones that provide the framework to build your DNA model. Remember, the strands are parallel, but begin from opposite directions.

6. Using the base sequence and Table 1, complete your suspect's DNA model.

7. Once everyone in the class has created their DNA models, compare your group's DNA model with those of your classmates. Are any two alike? Discuss your results.

8. When instructed by your teacher, open the sealed envelope. Analyze whether your suspect's DNA sequence matches the crime scene DNA sequence.

2. Arguing from evidence

a. Who was the person who took a bite from the apple on your teacher's desk?

b. Why is it important to correctly construct the DNA model of the different samples?

c. How could mistakes in the construction of the models lead to a mistake in the identity of the suspect?

3. Extending your investigation

A technique called DNA fingerprinting produces an image of patterns made by a person's DNA. Using an enzyme, scientists "cut" DNA strands in specific places. The DNA fragments are injected into a gel and an electric current is applied. As the fragments migrate across the gel, they create patterns (see image at right). Those patterns (DNA fingerprints) are related to the base sequences along the DNA strand. Suppose a serious crime has been committed. There are six suspects. Since blood was found at the crime scene, DNA fingerprints can be produced. Blood is drawn from the six suspects and DNA fingerprints are produced. By comparing the DNA fingerprints of the suspects to the blood from the crime scene, police can determine which suspects were present at the scene. Additional forensic evidence and witness testimony is needed to build a strong case against a suspect. Can you tell which suspects can be eliminated?

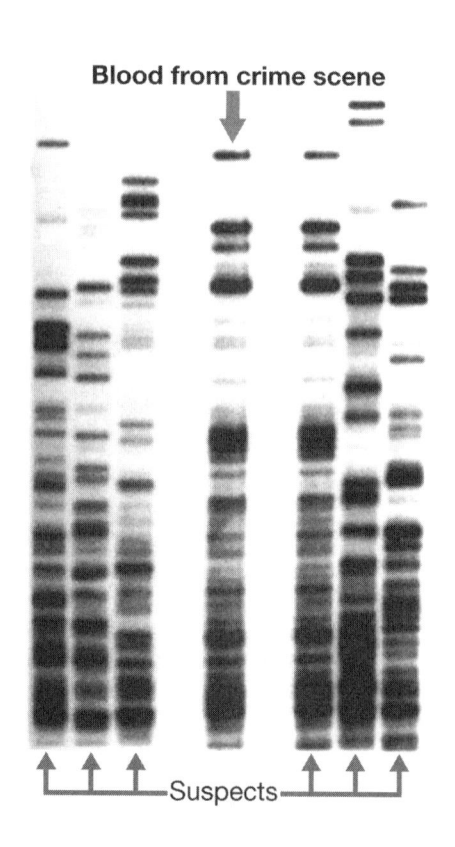

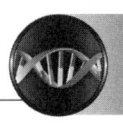

13A Crazy Adaptations

How can the environment influence traits?

When Darwin examined the finches on the Galapagos Islands, he noticed differences among different species from different islands. He noticed that the shape of finch beaks varied according to their primary food source. In this investigation, you will build a creature that is adapted to its environment and describe its adaptations. Then you will play the game of Adaptation Survivor.

Materials
- Crazy Traits kit
- Die

1. Determining your environment

Roll the die for each environmental variable and record your results in Table 1.

Table 1: Environmental data

Environmental variable	Possibilities with roll of the dice	Outcome
Surface color	1, 2 = Blue soil 3, 4 = Purple soil 5, 6 = Red soil	
Food source	1, 2 = chocolate candies 3, 4 = jumbo marshmallows 5, 6 = milkshakes	
Predator	1, 2 = *Hawkus giganticus* (flies over the land and snatches prey by their antennae) 3, 4 = *Frightus catus* (afraid of water but can run very fast) 5, 6 = *Microtus pesticus* (a blind army ant that crawls on the ground and attacks in large groups but cannot fly)	
Topography	1, 2 = flat 3, 4 = mountainous 5, 6 = swampy	

2. Developing a model

a. What is an adaptation? How do adaptations occur?

b. What kinds of adaptations would your creature need in order to survive in the environment in Table 1?

3. Choosing your traits

1. Think about adaptations that would help your organism survive in the environment you determined in Part 1. Circle the traits from Table 2 below that would be helpful adaptations.

Table 2: Possible genotypes and phenotypes of traits

	Trait	Genotypes and phenotypes
1.	Gender	XX - female XY - male
2.	Skin color	TT - red Tt - purple tt - blue
3.	Leg	TT - short Tt - short tt - long
4.	Foot	TT - webbed Tt - webbed tt - talons
5.	Arms	TT - long Tt - long tt - short
6.	Hands	TT - paws Tt - paws tt - claws
7.	Eye color	TT - red Tt - one red and one green tt - green
8.	Eyebrows	TT - unibrow Tt - unibrow tt - separate
9.	Beak	TT - trumpet Tt - trumpet tt - crusher
10.	Ears	TT - elephant Tt - elephant tt - mouse
11.	Antenna	TT - long Tt - long tt - short
12.	Antenna shape	TT - knob Tt - knob tt - star
13.	Tail	TT - long Tt - short tt - none
14.	Wings	TT - no wings Tt - no wings tt - wings

2. Complete Table 3 by filling in the genotype and phenotype for each trait you selected.
3. Tell whether each trait is an adaptation or not. If a trait is an adaptation, explain how it will help your creature survive in its environment. Record your answers in the last column of Table 3.

Table 3: Genotype and phenotype of your creature

	Trait	Genotype	Phenotype	Adaptation? (If yes, explain.)
1.	Gender			
2.	Skin color			
3.	Leg			
4.	Foot			
5.	Arms			
6.	Hands			
7.	Eye color			
8.	Eyebrows			
9.	Beak			
10.	Ears			
11.	Antenna			
12.	Antenna shape			
13.	Tail			
14.	Wings			

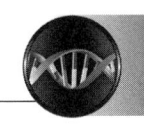

4. Using your model

a. Build your creature. List three adaptations you selected.

b. Describe your creature's adaptations for surviving in its environment. Make a table or draw your creature. Be creative!

c. Compare your environment and creature to the environments and creatures of the other groups in your class.

5. Playing the Game of Adaptation Survivor

Refer to Table 3 for the phenotypes of your creature. You will now see if your creature can survive the unpredictable conditions of a changing world. The object of the game is to be the last surviving creature in your environment. You earn or lose points based on whether your creature's particular set of traits are adaptations for survival. **When your creature earns a minus-three total points, it becomes extinct.**

1. Your teacher will choose someone to draw the first Environment Card and read it aloud. Each card describes two environmental conditions or events - read only the *one* condition that is facing you as you draw it. For each environmental condition, your creature may thrive (+1); be pushed closer to extinction (-1); or not be affected (+0).

   The scoring is based upon your creature's phenotype for a given trait. For example, if a successful land predator finds its way onto your island, your only chance for survival may be the set of wings that your creature possesses. This would give you an advantage over a non-winged creature. For that Environment Card you would earn a plus one (+1). If your creature does not possess wings, then you earn a minus one (-1).

2. There are some special cards in the deck, called Catastrophe Cards. If one of these cards is drawn, the outcome affects the class population in different ways.

3. Your class will continue drawing Environment Cards. Keep track of your group's points. Your teacher may ask you to explain your survival points.

4. Play until there is only one surviving creature. This is the winner of the game.
   NOTE: If a final card wipes out the last two or more creatures (complete extinction), then there is no winning survivor. Do you think that a complete extinction like this is likely to happen often? Why or why not?

13B Natural Selection

How do we model natural selection?

In the last investigation, you learned that there are favorable and unfavorable adaptations in a population. Favorable adaptations help an individual survive and are passed on to the next generation. In this investigation, you will simulate the process of natural selection. As you proceed, think about how a new species forms.

Materials

- 100 Red jelly beans (or paper punches)
- 100 Yellow jelly beans
- 100 Green jelly beans
- Poster sheet of red paper
- Poster sheet of yellow paper
- Poster sheet of green paper
- 3 Shallow dishes
- Plastic spoon

1. Creating a model

Imagine a population of tiny, candy creatures called *yum-yums* that live on Ketchup Island. Their main predator is a large, hungry creature called a *gobbler*. Gobblers can only open their eyes for one second at a time every 10 seconds. Yum-yums are the gobblers' favorite food because they taste so sweet.

Yum-yums normally have a red shell. But a mutation in the gene for shell color causes green- or yellow-shelled individuals. The percent of each shell color in the population is shown in Table 1.

Table 1: Yum-yum population on Ketchup Island

Shell color	Percent of population	Number of individuals
Red (R)	80	120
Yellow (Y)	10	15
Green (G)	10	15
Total population on Ketchup Island		150

1. In the simulation, you will use a large sheet of red paper to represent Ketchup Island. You will use colored jelly beans (or red, yellow, and green paper punches) to represent yum-yums. The gobbler can only use a spoon to collect its food supply. Work in groups of 3 or 6.
2. At the end of the investigation, your teacher MAY permit you to eat the yum-yums. If so, everyone in the group MUST wash and dry their hands. Handle yum-yums ONLY with a clean spoon.
3. Carefully spoon the correct number of yum-yums listed in Table 1 onto the red paper and spread them out randomly. DO NOT EAT ANY CANDY UNTIL YOU HAVE COMPLETED THE INVESTIGATION!

2. Using your model

a. Is there genetic variation in the yum-yum population? Explain your answer.

b. Why is genetic variation important in a population?

c. Explain why most of the yum-yum population is red. Use two of the following terms in your explanation: *adaptation*, *allele*, *phenotype*, *natural selection*, and *evolution*.

Investigation 13B Natural Selection

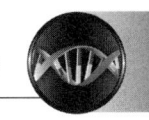

3. Modeling isolation

A huge storm has struck! The storm has blown 1/3 of the yum-yum population onto Mustard Island. It has blown another 1/3 of the population onto Relish Island. The remaining 1/3 of the population remains on Ketchup Island.

1. Remove 40 red, 5 yellow, and 5 green yum-yums from Ketchup Island using the spoon. Place them randomly onto a sheet of yellow paper (Mustard Island).
2. Remove another 40 red, 5 yellow, and 5 green yum-yums from Ketchup Island and place them randomly over a sheet of green paper (Relish Island).
3. You should now have 40 red, 5 yellow, and 5 green yum-yums remaining on Ketchup Island.
4. Complete the first row in Table 2. First, enter the number of individuals of each color on each Island. Then, calculate the percent of each shell color and enter in Table 2. Ketchup Island is done for you. To calculate percent, use the formula below:

$$\% \text{ of color} = \frac{\text{number of the color}}{\text{total number of jelly beans}} \times 100$$

Table 2: Yum-yum generations

Generations	Ketchup Is. Number of individuals	Ketchup Is. Percent of population	Mustard Is. Number of individuals	Mustard Is. Percent of population	Relish Is. Number of individuals	Relish Is. Percent of population
Parent	R = 40 Y = 5 G = 5	80 10 10	R Y G		R Y G	
1	R Y G		R Y G		R Y G	
2	R Y G		R Y G		R Y G	
3	R Y G		R Y G		R Y G	
4	R Y G		R Y G		R Y G	
5	R Y G		R Y G		R Y G	

4. Making a prediction

a. There are three steps to species formation: *isolation*, *adaptation*, and *differentiation*. Which step does Part 3 simulate?

b. What do you think will happen to yum-yum shell colors in each island population of yum-yums over time? Discuss ideas within your group, then make a prediction for each island.

5. Modeling natural selection

1. You will need six students in your group, two students for each island. For each island, one student is the *timer* and the other is the *gobbler*.
2. The gobbler sits in front of the island and closes her/his eyes.
3. The timer counts 10 seconds and then says: OPEN! The gobbler opens his eyes for one second, spoons the first yum-yum he sees, places it in a shallow dish, then closes his eyes again. This process is repeated 10 times.
4. Count the number of yum-yums left of each color, and enter in the next row of Table 2. Add up the total number of yum-yums and calculate the percent of each color and enter into Table 2.
5. Add 5 yum-yums of each color to your island. This step represents reproduction.
6. Repeat steps 2 through 5 for a total of 5 generations.
7. Enter the data for the other two islands from the other members of your group.

6. Analyzing the data

a. Make a *bar graph* of your data for each island. Plot generations (parent through 5) on the *x*-axis and percent of each color on the *y*-axis. Be sure to put a title on your graph and label the axes.

b. Compare the percentages of each shell color on each island for each generation. What changes, if any, occurred on each island over time? Does your data support your hypothesis from Part 4?

c. How does this investigation simulate natural selection?

d. Explain how populations on each island may become different species over time.

7. Extending your investigation

a. Yum-yums have a sweet taste that is preferred by gobblers. Suppose a mutation occurred in the gene for taste. The mutation resulted in offspring with a bitter taste. What would you expect to happen to that mutation over time? Explain what would happen to the percent of yum-yums with the mutation after many generations.

b. Make a list of other mutations in yum-yums that could result in organisms that are better adapted for survival. Be creative!

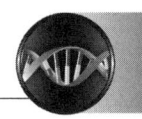

14A Relative Dating

How can you determine the sequence of past events?

Earth is very old and many of its features were formed long before people came along to study them. For that reason, studying Earth now is like detective work—using clues to uncover fascinating stories. The work of geologists and paleontologists is very much like the work of forensic scientists at a crime scene. In all three fields, the ability to put events in their proper order is the key to unraveling the hidden story.

Materials
- Construction paper
- Colored markers
- Tape
- Glue
- Scissors
- Different colors of modeling clay
- Different colors of sand or soil
- Rocks
- An empty shoe box or a clear container
- 2 Sheets of notebook paper

1 Sequencing recorded events

Carefully examine this illustration. It contains evidence of the following events:

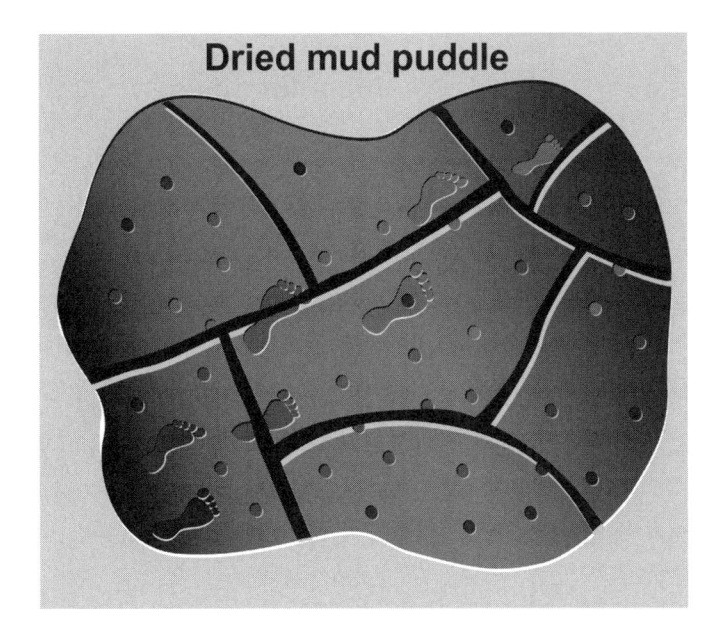

Dried mud puddle

- The baking heat of the sun caused cracks to formed in the dried mud puddle.
- A thunderstorm began.
- The mud puddle dried.
- A child ran through the mud puddle.
- Hailstones fell during the thunderstorm.

a. From the clues in the illustration, sequence the events listed above in the order in which they happened.

b. What tools or other scientific evidence could help us better explain what took place and when? Discuss answers with your class.

77

2. Evidence for the relative ages of rock formations

Relative dating is an earth science term that describes the set of principles and techniques used to sequence geologic events and determine the relative age of rock formations. Below are graphics that illustrate some of these basic principles used by geologists.

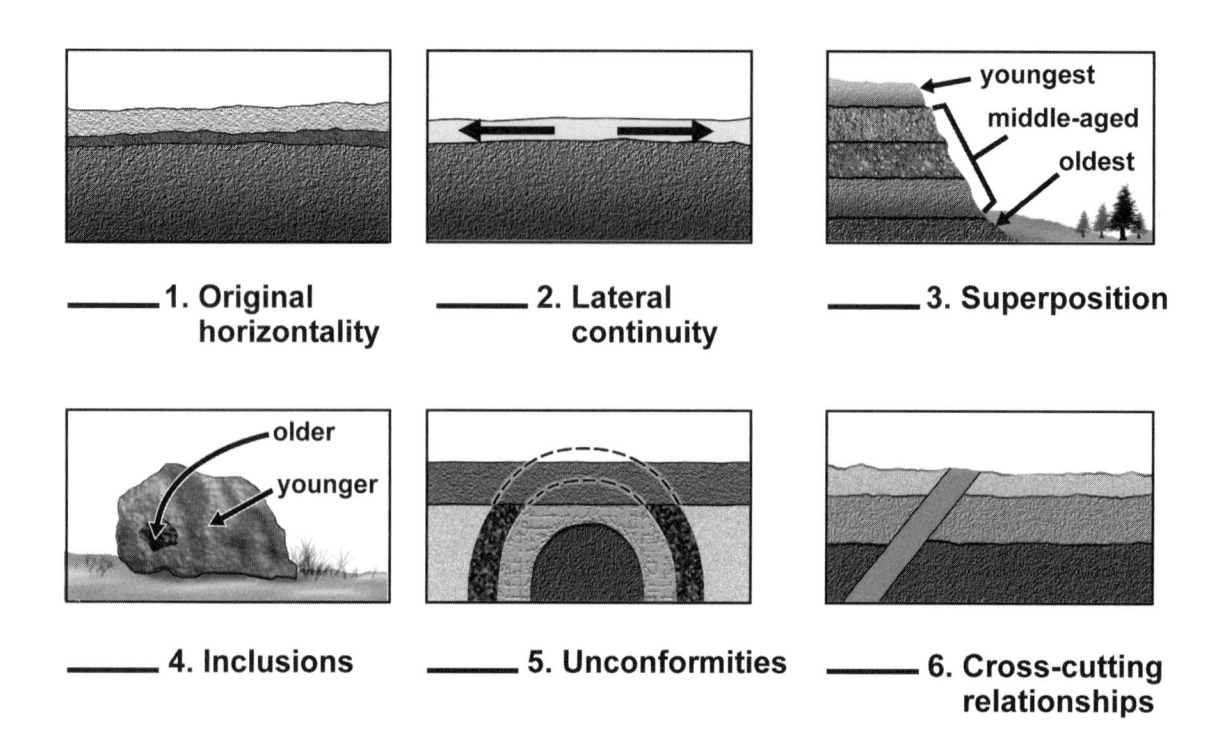

_____ 1. Original horizontality

_____ 2. Lateral continuity

_____ 3. Superposition

_____ 4. Inclusions

_____ 5. Unconformities

_____ 6. Cross-cutting relationships

Match each principle to its explanation. Write the letter of the explanation in the space provided under each graphic.

Explanations:

a. In undisturbed rock layers, the oldest layer is at the bottom and the youngest layer is at the top.

b. In some rock formations, layers or parts of layers may be missing. This is often due to **erosion**. Erosion by water or wind removes sediment from exposed surfaces. Erosion often leaves a new flat surface with some of the original material missing.

c. Sediments are originally deposited in horizontal layers.

d. Any feature that cuts across rock layers is younger than the layers it cuts through.

e. Sedimentary layers or lava flows extend sideways in all directions until they thin out or reach a barrier.

f. Any part of a previous rock layer, like a piece of stone, is older than the layer containing it.

Investigation 14A   *Relative Dating*

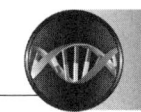

3 Sequencing events in a geologic cross-section

Understanding how a land formation, with its many layers of soil, was created begins with the same time-ordering process you used in Part 1. Geologists use logical thinking and geology principles like the ones described in Part 2 to determine the order of events for a geologic formation. Cross-sections of Earth, like the one shown below, are our best records of what natural events happened in the past.

Rock bodies in this cross-section are labeled A through H. One of these rock bodies is an **intrusion**. Intrusions occur when molten rock called *magma* penetrates into layers from below. The magma is always younger than the layers that it penetrates. Likewise, a fault is always younger than the layers that have faulted. A *fault* is a crack or break occurs across rock layers, and the term *faulting* is used to describe the occurrence of a fault. The broken layers may move so that one side of the fault is higher than the other. Faulted layers may also tilt.

a. Put the rock bodies illustrated below in order based on when they formed, from oldest to youngest.

b. Relative to the other rock bodies, when did the fault occur?

c. Compared with the formation of the rock bodies, when did the stream form? Justify your answer.

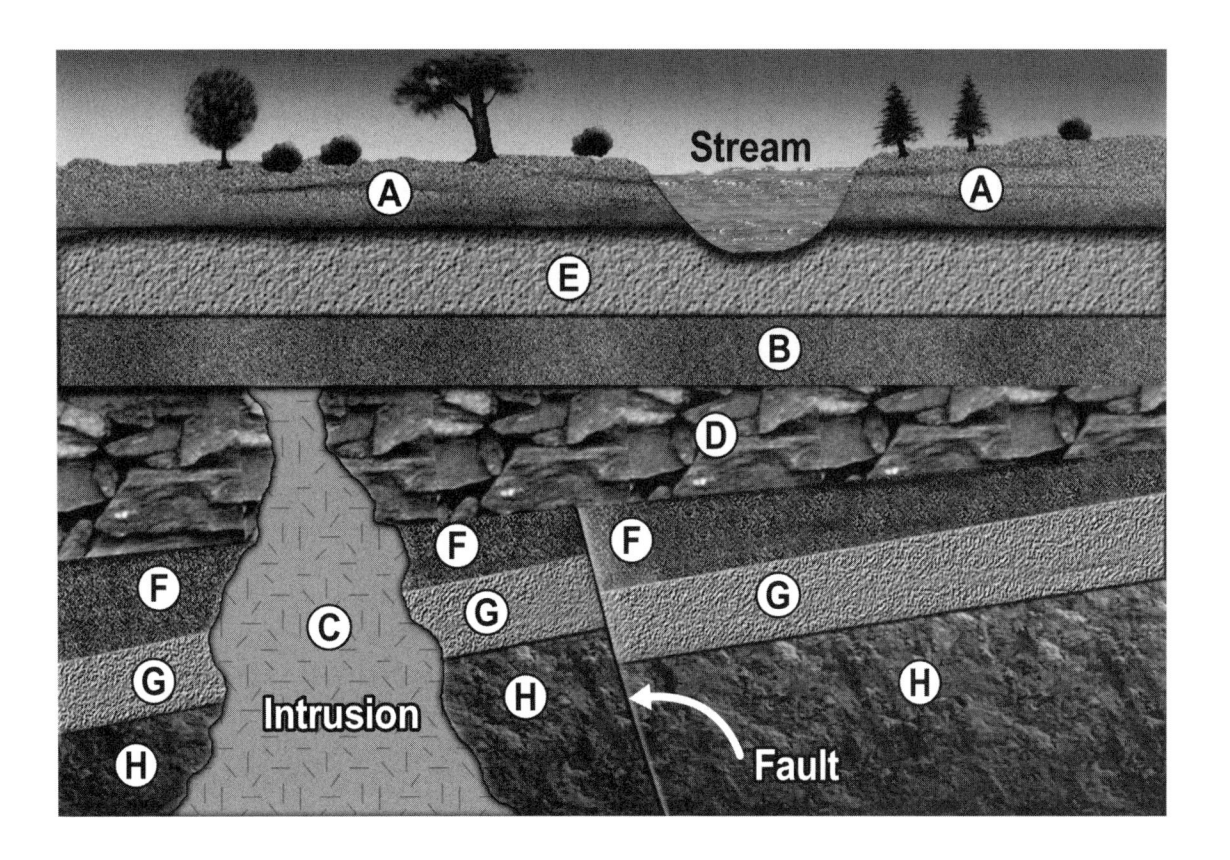

4. Creating clues for sequencing natural events

Now, your teacher will provide your group with some materials. Discuss ideas with your team for what past events took place, how to recreate them, and the evidence they might leave behind. Follow these guidelines in creating your "snapshot in time:"

- Set up a situation that includes clues that represent at least five events.
- Each of the five events must happen independently. In other words, two events cannot have happened at the same time.
- Use at least one geology principle that you learned in this investigation.
- Answer the questions below.

a. List your clues in order on one sheet of notebook paper. Include enough details in your clues so that someone can re-create the events that happened. Label this sheet "ANSWER" and include the names of your team members. On a second sheet, scramble the clues and tape this sheet to your model.

b. What relative dating principles are represented in your set of clues? Explain how these principles are represented.

c. Now, trade boxes and have a group of your classmates put your set of clues in order. When they are done, evaluate their work. How well did they guess your sequence? What did you fail to include in your model? If they made an error, explain what they missed.

5. Extending your investigation

a. In the investigation, you organized your thoughts using models. How does organizing your thoughts using models help you understand science?

b. Cross-sections like the illustration in Part 3 have been used to help explain amazing events, like the collision of continents or an ancient earthquake. If two continents collided, what features might you see in a cross-section of land?

c. Read about forensic science on the Internet or in your local library. How is forensic science like life science? Write a short paragraph that compares and contrasts these two branches of science.

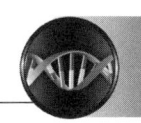

14B Interpreting the Fossil Record

How do scientists analyze the fossil record?

Fossils are the preserved remains of once-living things. After a living thing dies, it can be covered by wind- or water-carried rock particles. If undisturbed for a long time, these particles can be changed by pressure into sedimentary rock. In this way, parts of the once-living thing can be preserved in rock as a fossil. Fossils are like snapshots in an old photo album handed down in a family. Unfortunately, time and events have scrambled the album. Some pages are missing and others are out of order. Paleontologists are scientists who interpret the fossil record of past living things. By studying fossils, they are able to trace how living things appeared and changed over time. In this investigation, you will see that mathematics is essential to the interpretation of fossils.

Materials
- Copies of investigation handouts
- Metric ruler
- Scissors
- 4 Plastic sandwich bags
- Pen or pencil

1. Measuring fossils

To explore the way paleontologists study fossils, we will use the Common Atlantic Slipper as an example. Mollusks such as clams are ancient animals. They have been on Earth for hundreds of millions of years. The Common Atlantic Slipper is sometimes called a "living fossil," because both fossils and living slippers are still found with seaside rocks today. Slippers are mollusks with only one shell. They hold on to solid surfaces with their muscular foot, and the single shell forms a protective shield over their backs. Here is an actual-size picture of the belly (*ventral*) and back (*dorsal*) surface of a Common Atlantic Slipper:

A Common Atlantic Slipper
(actual size)

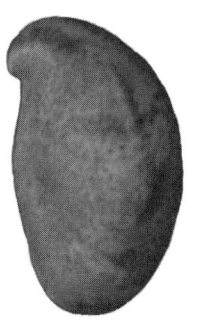

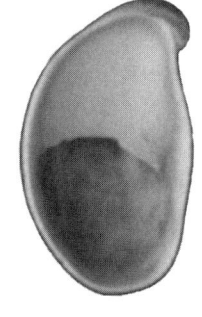

Dorsal side   Ventral side

When studying fossils, some characteristics provide better data than others. You might measure the *length* of the Slipper's shell. Or you might measure the *roundness* of the Slipper's shell. Then you would compare fossil samples from different times to see if that measurement changed as time passed.

Measuring the roundness of the shell provides more reliable data than measuring the length of the shell. Can you see why? Imagine that fossil shells were collected from a time when food was scarce. Scarce food means slower growth. Why might scarce food have a greater effect on the length of the shells?

2. Measuring shape

To see how scientists measure the roundness of Slipper shells, use the photograph in Part 1 and follow the steps below:

1. Measure the shell length along its longest dimension.
2. Turn your ruler ninety degrees and measure the shell width across the widest point.
3. Divide the width by the length.
4. The result will be a number between zero and one, called the shape index.

Shape Index

```
0      0.2      0.4      0.6      0.8      1.0
|<----------------------------------------------->|
Narrow                                        Round
```

Example:

$$Shape = \frac{Width}{Length}$$

$$Shape = \frac{24\ mm}{40\ mm} = 0.6$$

a. What is the shape index value for the Common Atlantic Slipper in the photograph in Part 1? What does the number tell you about its shape?

b. Which shell is more round, one with a shape index of 0.4 or one with a shape index of 0.5?

3. Representing measurements in a sample of Slipper shells

Imagine that you are a million years in the future. While digging in the side of a hill, you find the bones of an NBA basketball player. These bones belonged to a human who was much taller than earlier humans. You conclude that something caused humans to suddenly change so that they became very tall. Back in the present, we know that this conclusion is not correct. Humans did not suddenly become over six feet tall. What caused this error? What could you do to avoid this error?

The human height error was caused by having only one set of bones to measure. Look at your classmates around you. Would it be reasonable to pick a classmate and declare that everyone in your class stood that tall? Clearly many individuals must be measured. Then all of the measurements must be mathematically changed into a single height that represents your class.

Investigation 14B *Interpreting the Fossil Record*

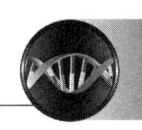

How do we change many class heights into a single height? This process belongs to the branch of mathematics called *statistics*. We will look at two statistical ways to represent a group of measurements, as a *graph* and as a *number*.

Representing sample measurements as a graph

A graph is a picture of all of the members of a fossil sample. There are 32 individual shells in this sample. They range from narrow shells with a shape index of 0.4 to more rounded shells with a shape index of 0.7. The graph allows you to see the number of each shape in the sample.

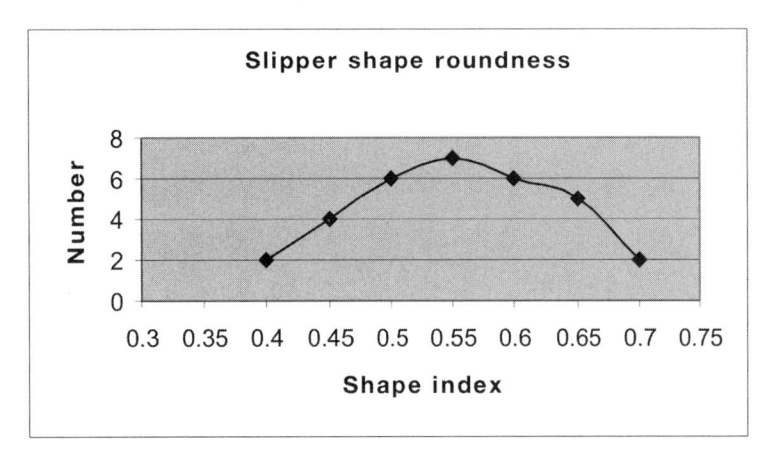

Slipper shape roundness

(Graph: Number vs. Shape index. Points: (0.4, 2), (0.45, 4), (0.5, 6), (0.55, 7), (0.6, 6), (0.65, 5), (0.7, 2))

a. How many shells have a shape index of 0.6?

b. What shape index number is the most common?

Representing sample measurements as a number

Showing measurements as a graph gives a lot of information about a single sample. But graphs can be difficult when you want to compare many samples, each with many measurements. Comparing many samples is much easier if the measurements in each sample can be converted into a single number. Then comparing the samples is as easy as comparing their numbers. There are many different statistical methods for changing a group of measurements into a single number, but we will use an easy one that you may already know, averaging.

1. Finding an average is easy. You add all of the measurements and divide the sum by the total number of measurements.

$$Average = \frac{Sum\ of\ measurements}{Total\ number\ of\ measurements}$$

2. The graph above shows the shape index for all thirty-two Slipper shells in a sample. Use the formula in step 1 to calculate the average shape index for this sample. The numerical data for the sample is given in Table 1.

Table 1: Shape index for a sample of Common Atlantic Slippers

Shell no.	Shape	Shell no.	Shape	Shell no.	Shape	Shell no.	Shape
1	0.4	9	0.5	17	0.55	25	0.5
2	0.45	10	0.6	18	0.65	26	0.55
3	0.55	11	0.55	19	0.5	27	0.55
4	0.6	12	0.65	20	0.55	28	0.7
5	0.5	13	0.6	21	0.65	29	0.6
6	0.5	14	0.4	22	0.5	30	0.45
7	0.45	15	0.65	23	0.7	31	0.65
8	0.6	16	0.55	24	0.6	32	0.45

4. Stop and think

a. Why is roundness used as a measurement instead of length for analyzing Slipper shell fossils?

b. Why is a graph a good way to compare the shape index of all of the Slipper shells in a single sample?

c. Why do we use an average value for comparing the shape index of many Slipper shell samples?

5. Interpreting the fossil record

Congratulations! You now have all the skills needed to interpret imaginary fossil records. At the beginning, we said that fossils are like a picture album of life with missing and out of order pages. Your task is to reconstruct that picture album and interpret the changes that took place. Paleontologists have a specific way that they represent each album page. It looks like the graph shown to the right.

This graph is unusual in that time is shown on the left side. Usually time is shown along the bottom. Paleontologists use this method because it reflects the way they find fossils. Older fossils are normally deeper in the ground than younger fossils if the area is undisturbed.

The oldest fossils are very old! Scientists who deal with deep time use the *geological time scale* to represent large amounts of time. The divisions of the scale have names that correspond to millions of years ago, abbreviated *mya*.

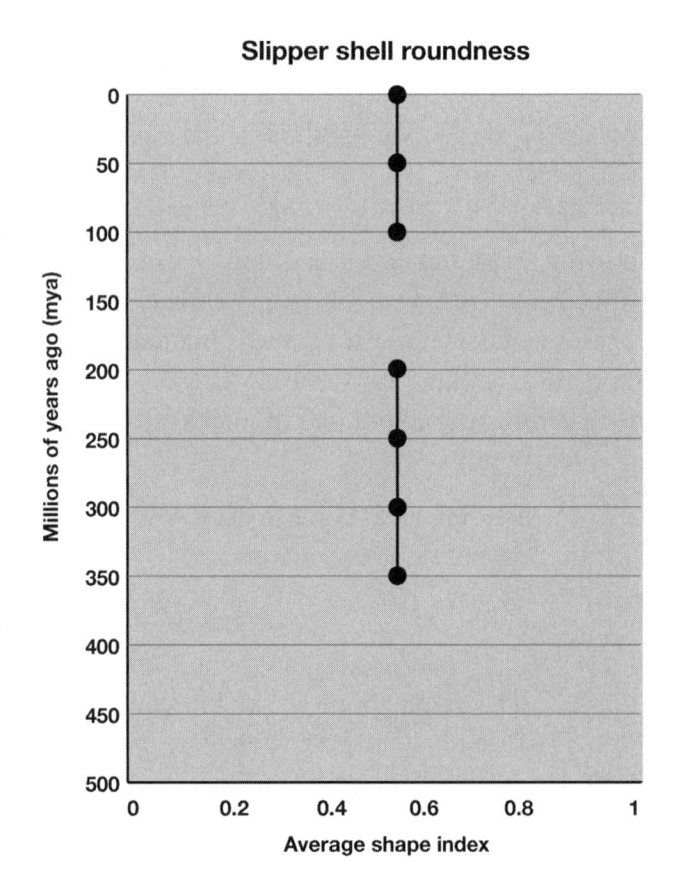

Investigation 14B *Interpreting the Fossil Record*

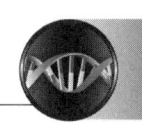

This fossil record shows the plotted data from seven samples of our Common Atlantic Slipper. Here are the original data that were used to prepare the graph:

Table 2: Average Slipper shape index for samples

Millions of years ago (mya)	Shape index	Millions of years ago (mya)	Shape index
0 (Now)	0.55	300	0.54
50	0.56	350	0.55
100	0.55	400	no sample
150	no sample	450	no sample
200	0.55	500	no sample
250	0.56		

Now we can make some interpretations about the history of the Common Atlantic Slipper. Here are some questions to guide you. You should be able to give specific reasons for your answers. Write down your answers before you read the paragraph below. Then compare your answers to the ones given.

a. Is there evidence for change in the shape of the Slipper, or has its shape been the same through time?

b. Is the Common Atlantic Slipper extinct?

c. There is no sample from 150 million years ago. Where on the graph would you expect the missing point to appear? Explain your answer.

d. When did the Common Atlantic Slipper first appear?

The first three questions are easy to answer. The shape has been constant through time as shown by the straight vertical line segments. The vertical line segments are due to the almost unchanging shape index of each sample. The Slipper is not extinct because we have a present-day sample. The missing sample would probably appear in line with the before and after samples because the shape has been constant through time. But the fourth question is tricky. We might be tempted to conclude that the lack of samples before 350 mya means that the Slipper appeared sometime between 400 and 350 mya. This is a risky conclusion, however. Why? Can you be sure that there are no older Slipper fossils?

6. Investigating an imaginary fossil record

You will work with a partner for the rest of the investigation.

1. Prepare a fossil record sheet by taping together the two halves provided by your teacher.

2. Use scissors to cut out all of the squares on the data page. Each square shows a BOLD Set Letter, the geologic time in mya, and the average shape index for the sample.

   ```
   B        — Set letter
   300 mya  — Geologic time
   SI 0.54  — Average shape index for sample
   ```

3. Separate the squares into four sets by the Set Letter on each square. Place each set in a separate plastic sandwich bag.

4. Place all of the squares from one set in their correctly plotted positions on the fossil record sheet that you prepared.

5. Study and discuss the line formed by the squares on your fossil record with your partner. Make a sketch of the line in your notebook and label it with the Set Letter. You probably know about graphs, but here are few ideas to keep in mind:

   • A straight vertical line indicates no change.
   • A straight tilted line indicates change at a constant rate.
   • A line that curves away from vertical indicates that change is speeding up.
   • A line that curves toward vertical indicates that change is slowing down.
   • A kink in a plotted line, called an inflection, has a cause. Something is different after the inflection point.

6. Record your interpretations of the fossil record for this sample set. Be sure to include your reasons for making these interpretations.

7. Repeat steps 4 - 6 for the next set. Continue until you have interpreted all four sets. Keep in mind that each set is completely separate from the others with its own set of interpretations.

8. At the end of the time allowed, your teacher will sketch the plot of each sample set on the board and ask teams to share their interpretations.

Investigation 14B *Interpreting the Fossil Record*

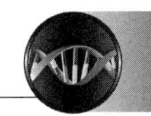

7. Applying your knowledge

a. **Data Set A**
    This data set is incomplete. Very few fossil records are complete, but paleontologists can often make strong interpretations even with some gaps. After you have interpreted Set A, rate your confidence in your interpretation as strong, fair, or weak.

b. **Data Set B**
    What is special about Set B after 250 mya? Use your imagination to create and describe an environment that might explain this. A good starting point might have to do with predators or food availability.

c. **Data Set C**
    Sometimes this is all that paleontologists have to work with. What interpretations can you make confidently? When you discuss Set C with your class, you can offer other possibilities, but for now, record only those that you can support with these data.

d. **Data Set D**
    Here is a good example of why data are displayed as graphs as well as numbers. This is a graph of all of Slipper shells from 250 mya in Data Set D. Although the average Shape Index is 0.54, the graph gives a hint about an important change in progress. Explain how this graph shape relates to the Data Set D twin lines after 250 mya. What biological event do you think produced the twin lines after this date?

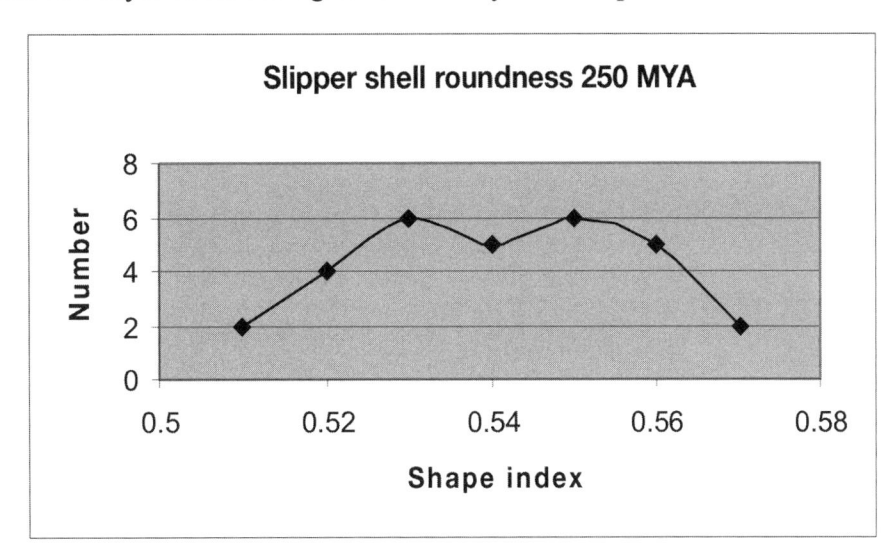

Slipper shell roundness 250 MYA

(Line graph: Number vs. Shape index. Points approximately at (0.5, 2), (0.52, 4), (0.53, 6), (0.54, 5), (0.55, 6), (0.56, 5), (0.57, 2).)

87

15A Creature Cladogram

What type of information can be used to create a cladogram?

Recall that <u>taxonomy</u> is the science of grouping living things on the basis of similar characteristics. Organisms are classified according to their structures and evolutionary relationships. Cladograms show these relationships. Organisms can have similar features, yet not be closely related at all. For example, wings are *appendages*, projecting parts of organisms with a distinct function. Wings are found on both birds and insects, but do not have the same evolutionary origin. Sometimes organisms that appear very different end up in the same group. Who would have imagined that whales are mammals that nurse their young like humans do? In this investigation, you will identify common characteristics in a group of imaginary creatures and use this information to create your own evolutionary tree.

Materials
- Creature Cards
- Large sheet of paper
- Colored pencils or markers

1. Creating a model

1. Examine the sheet of the imaginary creatures. Compare and contrast the major features of the creatures.
2. Make a list in Table 1 of the different features of the animals that may provide a way to classify them into groups.

Table 1: Imaginary creature features

Creature number	Features
1	
2	
3	
4	
5	
6	
7	
8	
9	
10	

Investigation 15A   *Creature Cladogram*

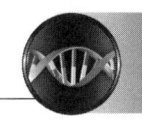

2. Developing your model

a. What types of features are contained in Table 1?

b. Are there any common characteristics that all ten creatures share?

3. Using your model

1. Fill in the table below as you try to determine the relationships among different creatures.

Table 2: Common features

Type of feature	Feature	Creature number
Number of appendages on body	Two	
	More than two	
Type of appendage on body or head	Fin	
	Leg	
	Tail	
	Tentacle	
	Wing	
Adaptation of head	Antenna	
	Beak	
	Ears	
	Hair	
	Horn	
Adaptation of appendage	Claws	
	Feet with toes	
	Flippers	
	Feathers	
	Suckers	
Adaptation of mouth	Forked tongue	
	Lips	
	Teeth	

2. Use the information in Table 2 to divide the organisms into different groups whose members are somehow related to each other. Try to reach some agreement in your group based on the evidence you observe.

4. Constructing explanations

a. Based on your answers in Table 2, which creatures are most closely related? List their numbers, then state which features they share that were also present in their common ancestor.

b. Which creatures are most distantly related? List their numbers and explain your group's reasoning.

c. Using your information in Table 2, create a *cladogram* that illustrates relationships among your creatures. An example cladogram is shown to the right. Your cladogram will look much different than the one shown. It should resemble a tree with two forked branches at each point, but it could have many paths. Be creative!

Draw your cladogram on separate paper.

d. Choose a point on your diagram where two organisms branch. Between the branches, be sure to list the characteristic feature shared by ALL the creatures after each branching point.

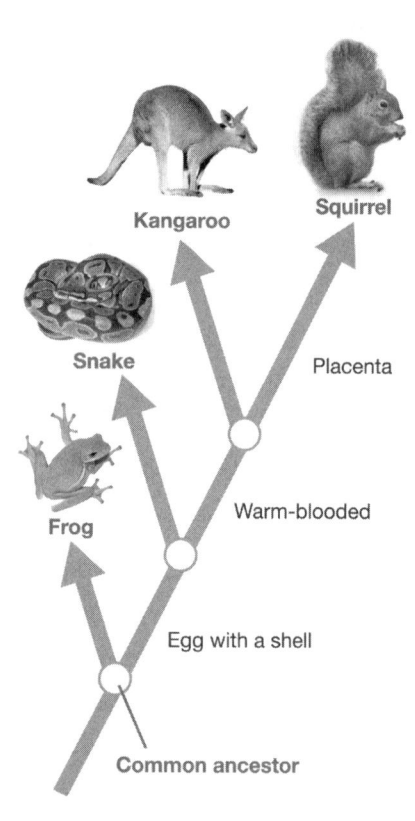

5. Extending your investigation

Once you have completed the investigation, pick one creature and using the information collected above, write a brief description of the creature's habitat. The goal of this activity is to be creative and use as much information as you can in describing the adaptations of this creature. Include things like where it lives, what it might eat, and so on.

Investigation 15B Bread Mold

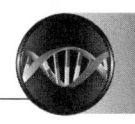

15B Bread Mold

What is mold and how does it grow on bread?

<u>Fungi</u> are found all around you. Examples include baker's yeast, mushrooms, and common bread mold. Mold grows from spores that can be found almost anywhere. The spores, under the right conditions, grow and give rise to the fuzzy organisms that can be observed on week-old bread left on the countertop. Most of us look at these organisms with disgust and quickly throw the bread in the trash! In this investigation, you will grow bread mold and examine the colonies.

Materials
- Homemade bread or tortilla (without preservatives)
- Paper towels
- Distilled water
- Plastic bags
- Marker
- Compound or digital microscope (400x)
- Microscope slides and coverslips
- Dropping pipette
- Tweezers

1. Providing conditions for growth

1. Remove a single slice of bread and place it in a plastic bag.
2. Wet a piece of paper towel with distilled water, and place the wet paper towel into the plastic bag.
3. Secure the plastic bag containing the paper towel and the bread.
4. Label the plastic bag with your initials.
5. Place the bag in a designated area of the room. Wash and dry your hands thoroughly.

2. Making predictions

a. What type(s) of organism(s) do you think will grow on the bread?

b. How long do you think it will take for organisms to appear?

c. What conditions do you think are needed to produce bread mold?

3. Making observations

1. Examine the bread through the plastic bag.
2. Record your observations in Table 1.
3. Repeat Steps 1 and 2 daily until you begin to see fuzzy threads or black areas growing on the bread.

Table 1: Bread mold observations

Date	Observations

4. Once you observe organisms without magnification, you may want to look at these organisms with your microscope. Use low power (4x) on a compound or digital microscope. Scan for changes on the surface of the bread. Proceed to step 5 only when you observe black, fuzzy areas.
5. Place a single drop of water on a clean microscope slide.
6. Using the tweezers, remove a small sample of the black bread mold and place it into the drop of water.
7. Gently place a cover slip on top of the drop of water, and examine the slide under your microscope. Remember, always look at your sample under low power first and then move to a high-power objective.
8. In Table 2, make a sketch of your observations. Label any structures you can identify.

Table 2: Microscopic evaluation of mold growth

Date	Observations

9. When finished, thoroughly wash and dry your hands.

4. Constructing explanations

a. Compare observations you made in Table 1 with those in Table 2. What differences did you notice when you changed the scale of your observations?

b. What type of organisms did you observe growing on your bread?

c. Why is it important to use homemade bread?

d. What is the best way to keep foods free from mold?

e. What other foods do you think would grow mold over time?

5. Extending your investigation

This investigation can be used as a stepping stone for asking more questions about mold growth. For example, does mold grow better in the dark or light? Does a certain type of bread work better than others? Design a controlled experiment that explores one of these questions, or think of other questions you and your classmates could explore. Before experimenting, always obtain permission and conduct all procedures with your teacher's supervision. Experimenting with microorganisms can be dangerous to you or others.

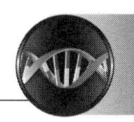

16A Leaf Structure and Function

How do different parts of a leaf function together?

In most vascular plants, leaves are the principal organs for photosynthesis. Although leaves vary in their shapes and sizes, most have a thin, flat blade and veins. Some of the variation in leaf structure is due to habitat. For example, the colored, fleshy leaves of *T. zebrina* evolved differently than the long and thin needles of a pine tree. Leaf cells and tissues are highly specialized to work with a plant's vascular system. Much like a human's circulatory system, a plant's vascular system moves materials to and from leaves. In this investigation, you will identify the structures in a leaf and learn how they function together.

Materials

- Prepared slide of a leaf cross-section
- Digital or compound light microscope (400x)
- Leaf from a *T. zebrina* plant
- Microscope slide and coverslip
- Forceps
- Salt solution (1%)
- Dropping pipette
- Textbook

1. Leaf structure and function

1. Obtain a microscope and prepared slide from your teacher.
2. Examine the slide on low power (4x). Center the tissue in the field of view and switch to medium power (10x). As you work you may need to change the magnification or move the slide carefully.
3. Locate and identify the structures using the graphic below as a guide:

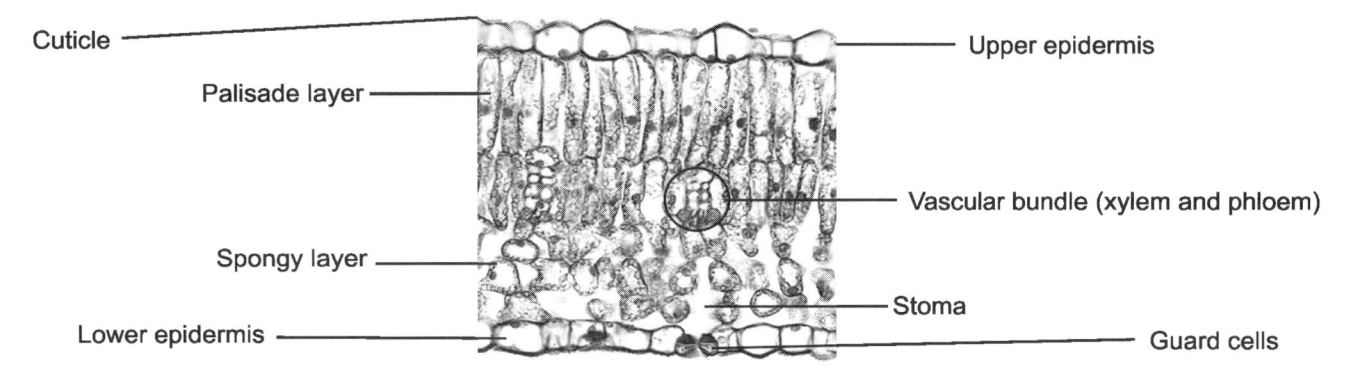

Labels: Cuticle, Palisade layer, Spongy layer, Lower epidermis, Upper epidermis, Vascular bundle (xylem and phloem), Stoma, Guard cells

4. Complete Table 1 by naming the leaf structure next to its function. You may use your textbook.

Table 1: Matching structures to their functions

Leaf structure	Function
1.	A. Light can easily pass through this layer
2.	B. Open and close the stomata
3.	C. Where most of the photosynthesis takes place
4.	D. Transports carbon compounds like sugars
5.	E. Waxy layer that protects the leaf surface
6.	F. Pore that allows carbon dioxide to pass into the leaf
7.	G. Transports water and dissolved nutrients
8.	H. Many spaces between cells that allow carbon dioxide to pass through

2. Using your model

a. Most of the structures in a leaf evolved to carry out a specific process to survive. What is that process called?

b. Which cell organelles are the key to this process? Where does the energy these organelles need originally come from?

c. Recall the importance of cellular respiration to animals. The cycling of matter between plants and animals is key to the survival of both. Name two substances plants produce that are needed by animals. Which substance does an animal release that can enter a plant's stomata to be cycled again?

3. Observing stomata

1. Remove a fresh leaf from a *T. zebrina* plant. Examine it closely. The lower epidermis is purple, while the upper epidermis is silver-and-green striped.
2. Gently snap the leaf in half, purple sides touching. With your fingernail or forceps, peel a portion of the lower epidermis from the leaf, starting at the cut edge.
3. Place a drop of water on the slide. Gently lower the peel, outer side up, in the drop of water. Try to lay the peel flat on the microscope slide. Wrinkled portions have too many layers of cells and tend to trap air bubbles. Place a coverslip on top of the peel.
4. Observe your slide with your microscope. Locate the guard cells under low power (4x).
5. When you find the guard cells, observe a stoma under medium power (10x).

4. Constructing explanations

a. When guard cells swell with water, they look like fat sausages. This closes the stomata. When guard cells lose water and shrink, the stomata open. Are any of the stomata open?

b. Recall from your observation of the prepared slide of a leaf that a stoma opens into an air space of the spongy layer. How does this arrangement help the process of photosynthesis?

c. Stomata are also critical for transpiration (water movement) in plants. What would happen if a plant could not close its stomata?

5. Arguing from evidence

Can you see chloroplasts in the guard cells? Scientists do not yet know why. Choose one position below to argue and give one example of evidence that supports your argument:

(1) Chloroplasts trigger the guard cells to open.

(2) Chloroplasts produce energy for the guard cells.

(3) Chloroplasts make starch for the guard cells.

6. Extending your investigation

1. Remove the coverslip. Use a tissue to remove as much of the water as you can. Now add a drop of 1% salt solution.
2. Place the coverslip onto the leaf and examine the stomata under medium power (10x).

a. Did the appearance of the stomata change? How did they change?

b. What caused the change in appearance of the stomata? (Hint: Review osmosis in Chapter 8 of your text.)

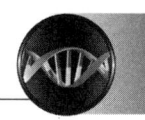

16B Flower Dissection

How does the design of a flower help in its pollination?

Do you know where the saying "the birds and the bees" came from? It all started with flowers. Plants require pollinators to carry their pollen to fertilize other flowers. Without the pollen, flowers would not be able to create seeds for reproducing. A number of different animals can pollinate flowers. Bats and Luna moths are attracted by odor. Monarch butterflies choose orange, red, or yellow flowers. Bees see ultraviolet light and prefer minty odors, but cannot see the color red. Flowering plants evolved specialized structures to increase their probability of successful reproduction.

Materials

- Large flower
- Newspaper
- Forceps
- Cellophane tape
- Hand lens
- Metric ruler
- Digital or compound microscope (400x)
- Glass slide and coverslip
- Glucose solution (1%)
- Dropping pipette

1. Making observations

A flower's only purpose is reproduction for the plant. In addition to male and female parts, flowers have petals and sepals.

1. Open a sheet of newspaper. This will be your work area. Lay the flower on the newspaper.
2. Review the parts listed on the flower diagram below. Pay close attention to the color, odor, shape, and texture of the flower.

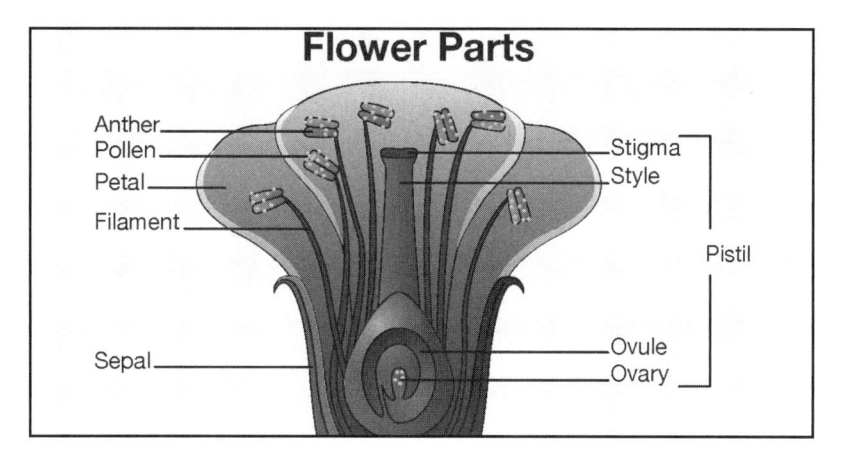

Flower Parts

(Diagram labels: Anther, Pollen, Petal, Filament, Sepal, Stigma, Style, Pistil, Ovule, Ovary)

2. Making inferences

a. Consider your flower's design. List one adaptation, suggest a pollinator that might be attracted to it, and explain why.

b. Some plants have flowers with no petals at all. Male and female parts are located on separate flowers, but no animals pollinate them. How might these plants get fertilized?

c. Imagine every different plant species all had blue flower petals and all animal pollinators could see that blue color. Would every plant species get fertilized and successfully reproduce? Why or why not?

3. Evidence of adaptations

1. Sepals are flower parts that cover and protect the flower bud. Record the number and color of the sepals in Table 1.
2. Gently pull the sepals away from the flower using forceps, being careful not to damage the structures underneath. Tape down a sepal onto the area labeled Flower Parts or in your science journal. Label it.
3. Record the number and color of the petals in Table 1.
4. Gently pull back each of the petals revealing the male and female parts inside. Carefully remove and tape down a petal. Before taping, be sure to separate any other structures that may be attached to the petal.

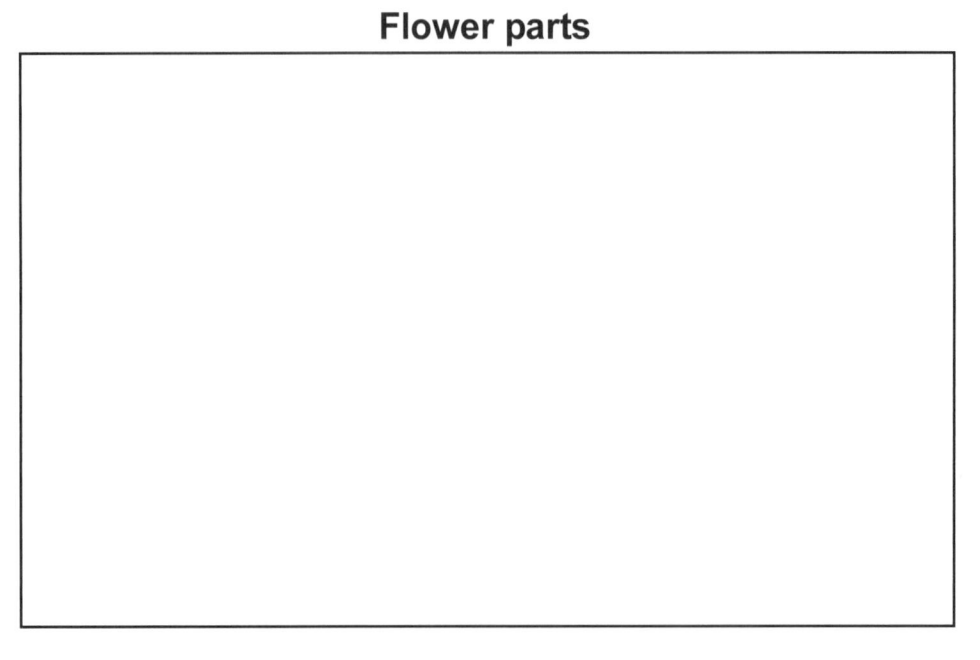

Flower parts

4. Examining the male parts of the flower

1. Use the forceps to remove all of the stamens. Record the number and color of the stamens in Table 1.
2. At random, select three unbroken stamens. Measure the length, in mm, of each of the three stamens. Record your results in Table 2 and calculate the average. Tape down one stamen with an anther. Label them.
3. Obtain a slide. Tap the anther over the slide so that the pollen from the anther falls onto the slide. Add a drop of 1% glucose solution to the slide and place a coverslip over the water and pollen.
4. Place the pollen slide on the microscope stage, and observe the pollen grain under low power (4x), medium power (10x), and high power (40x). Make a detailed sketch of one pollen grain at the highest power.

Investigation 16B   *Flower Dissection*

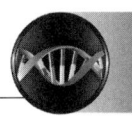

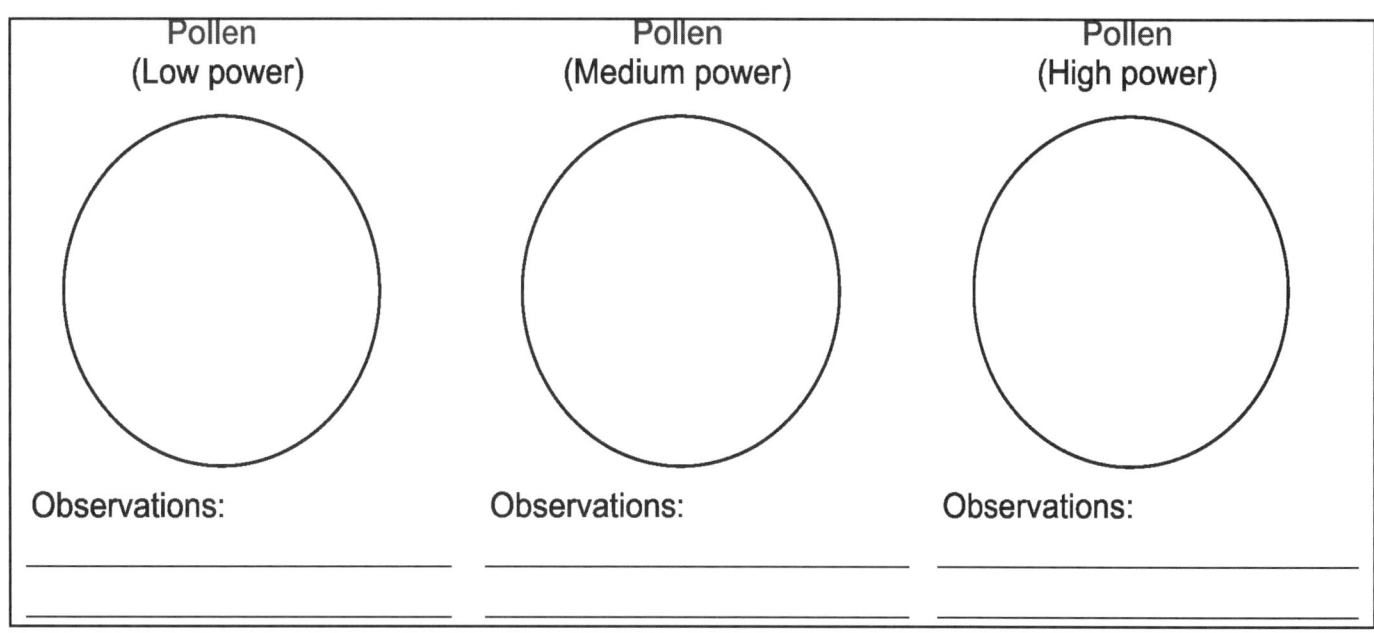

Pollen (Low power)	Pollen (Medium power)	Pollen (High power)
Observations:	Observations:	Observations:

5. Examining the female parts of the flower

1. Record the number and color of the pistil(s) in Table 1.
2. Measure the length, in mm, of the pistil and record your results in Table 2.
3. Examine the overall appearance of the pistil and note any unique qualities. Firmly hold the pistil flat on the newspaper.
4. Your teacher will demonstrate how to reveal the ovules inside the ovary.
5. Use the hand lens to examine the ovary. Next to your taped flower parts, sketch and label a female pistil with stigma, style, ovary, and ovules.

Table 1: Number and color of flower structures

Flower Structure	Number	Color
Sepal		
Petal		
Stamen		
Anther		
Pistil		

Table 2: Length of the stamen and pistil (mm)

Flower	#1	#2	#3	Average
Stamen				
Pistil				

97

6. Constructing explanations

a. Do you think all flowers have the same pollen shape? Why or why not?

b. What is the benefit to the plant if the flower attracts a pollinator? More than one pollinator?

c. Many flowers have several stamens and only one pistil. What might be the purpose of this design?

d. Compare the length of your pistil to the average length of the stamen. Why would one be longer than the other?

e. Which adaptations of your flower (color, odor, size, pollen shape, or arrangement of parts) help to increase the probability of successful reproduction? Explain.

7. Extending your investigation

- After the female parts of the flower are pollinated, it may develop into fruit. Obtain a piece of fruit. Try to match parts of the fruit to parts of the flower from which they developed.

Or

- Different flowers are designed for different pollinators. For example, some flowers are pollinated by bees. Others are pollinated by the wind. Would a flower have any major differences because of the type of pollinator? Design an experiment to find the answer. Check in with your teacher before carrying out your investigation.

Investigation 17A   *Observing Planarians*

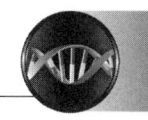

17A Observing Planarians

How do behaviors enhance an animal's probability of survival?

Planarians are invertebrate animals belonging to the *Phylum Platyhelminthes* (flatworms). You can easily find planarians on your own. Shake pond weeds into a pan or turn over stream rocks and look carefully at the rock surfaces. You can also collect flatworms on your own. Put a small pellet of canned pet food in an old nylon stocking. Secure that "bag" in a streambed or pond shore overnight. In the morning you may find a collection of flatworms crawling over the bag! In this investigation, you will observe planarians and identify their structures and behavior. You will also design and conduct a few experiments.

Materials
- Live planarians (*Dugesia*)
- Textbook
- Petri dish (60 mm x 15 mm)
- Spring water
- Dropping pipette
- Artist paintbrush
- Digital or compound microscope (400x)
- Depression slide and cover slip
- Food for planarians
- Black construction paper (10 cm x 5 cm)
- Metric ruler

1 Observing your planarian

1. Obtain a Petri dish containing a freshwater planarian, also known as *Dugesia*. Record all of your answers and data for Part 1 in Table 1.
2. List 3 characteristics of flatworms.
3. What type of *symmetry* does this worm have? The term symmetry refers to the body plan of the organism. Use the diagram below as a guide.

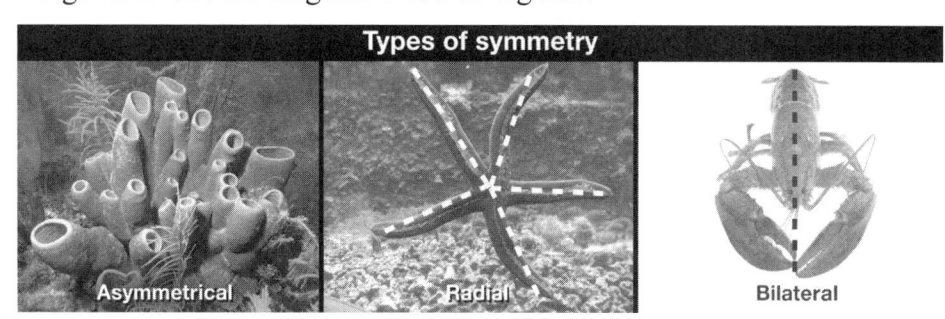

Types of symmetry

Asymmetrical   Radial   Bilateral

4. Using the dropping pipette, carefully move your planarian into a depression slide containing a few drops of spring water. The paintbrush can help you gently move the worm off surfaces.
5. Observe your worm using a microscope on low power (4x). Sketch the planarian. Label the eyespots. Label the *anterior* (head) and *posterior* (tail) ends.
6. Carefully transfer your worm back to its petri dish.
7. Measure your planarian. You can do this by placing a metric ruler under the field of view of the microscope or using the measuring feature on an image created by a digital microscope. Estimate the length of the worm in millimeters. Carefully transfer your flatworm back to its Petri dish.
8. Share your data with the class. When all the lengths are recorded, determine the average planarian length.

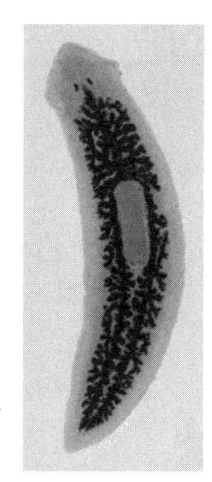

99

Table 1: Planarian facts and observations

Characteristics of flatworms: a. b. c.	
Type of symmetry:	
Sketch:	
Length of your planarian (mm):	Average length of planarians (mm):

2. Observing animal behavior

Observe the planarian for five minutes. Recall that all living things can respond to a stimulus. Flatworms have sensory receptors in a simple ladder-like nervous system. Two ganglia located in the head coordinate response. You will observe the response of the planarian to several stimuli. Record your observations in Table 2.

a. Is it active or still?
b. The planarian has a muscular system running across and down its body. What evidence do you see of muscles working together?
c. Where in the dish does your planarian spend most of its time?
d. Use the paintbrush to gently poke the planarian. How does it respond to touch?
e. Make a current in the water with a pipette. How does the planarian react?
f. How does it respond to the light of the microscope?
g. Your teacher will give you a piece of food for your planarian. Drop the piece of food into the Petri dish with the planarian. Observe the planarian's reactions. It may take a few minutes. How does it respond to food?

Table 2: Planarian movement and behavior

Movement	a. b.
Location preference	c.
Response to stimulus: d. touch e. current f. light g. food	d. e. f. g.

h. Which system is directly involved in taking in food?
i. Where in the body is food taken in?

Investigation 17A *Observing Planarians*

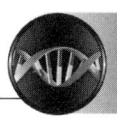

j. What is the name of the organ used for feeding in the planarian? You may look this up in your textbook.

Planarians display the behavior of being right- or left-handed. You can discover whether your worm is right- or left-handed by flipping the planarian over on its *dorsal* surface (back) and seeing which direction it flips to turn back over. If it rolls to the right, it is right-handed; if it rolls to the left, it is left-handed. Do five trials to determine the handedness of your planarian. Fill out Table 3. Compare your results to others in your class.

Table 3: Planarian "handedness"

Trial 1	
Trial 2	
Trial 3	
Trial 4	
Trial 5	
Based on your trials, is your planarian right- or left-handed?	

3. Exploring planarian reproduction

a. Planarians are hermaphrodites. How does a being a hermaphrodite increase the probablility of successful reproduction?

b. Planarians can also reproduce by regeneration. Is this method of reproduction sexual or asexual?.

c. Which cell process occurs to replace tissues?

If time permits, you can explore how planarians reproduce sexually and asexually. To do the experiment:

1. Ask your teacher for another planarian and place it on a glass slide with a drop or two of water.

2. Your teacher will demonstrate or perform the necessary cut. Observe the two pieces of the planarian under the microscope.

3. Place the pieces in a Petri dish with spring water and cover. Label the lid with your NAME and DATE and TIME.

a. Make a prediction: How long do you think (in days) it will take for your planarian to completely regenerate (replace) its organs and tissues?

b. What cell process occurs to replace these cells?

c. How does regeneration increase the probability than a flatworm will successfully reproduce?

4. Designing a behavior experiment

a. Design an experiment to test which condition the planarian consistently prefers: light or dark. First, state your hypothesis.

b. Describe your design for the experiment.

c. Get you teacher's permission to conduct your experiment to determine whether the planarian prefers light or dark. Construct a data table and record your observations and data.

d. Write your conclusions. Make sure you state whether your conclusion supports or refutes your hypothesis. Include your reasoning.

17B The Mammalian Eye

What are the structures of the mammalian eye and how do they function together?

The mammalian eye is an amazing organ. It consists of many specialized cells and tissues that make up several different structures. The structures have certain functions and, together, they form images that are interpreted by the brain. In this investigation, you will identify the most important structures and tissues in a sheep's eye and learn their functions.

Materials
- Preserved sheep's eye
- Dissection forceps
- Dissection probes
- Dissection scissors
- Dissection tray
- Protective gloves
- Apron
- Goggles
- Paper towels

Safety Note: Wear gloves, apron, and safety goggles during the investigation.

WARNING — This lab contains chemicals that may be harmful if misused. Read cautions on individual containers carefully. Not to be used by children except under adult supervision.

1 Planning your investigation

1. Put on your gloves, apron, and safety goggles.
2. Your teacher will give your group a preserved sheep's eye in a dissection tray. Handle the sheep's eye with care. The chemicals used to preserve it may be toxic.
3. Obtain a dissecting kit containing dissection probes, scissors, and forceps. Handle the equipment with care and do not play with it.

2 External features of the eye

1. Examine the outside layer of the sheep's eye. Notice the yellow tissue surrounding the eye. This is fat tissue. Using scissors, trim any excess fat tissue from around the eye.
2. Locate the sclera, the cornea, and the optic nerve.
3. The *sclera* is the tough outer coating of the eye. The sclera gives the eye its shape and helps to support and protect the delicate inner parts.
4. The *cornea* is the transparent front part of the eye. Together with the lens, the cornea refracts light and helps the eye to focus. The cornea refracts the light rays entering the eye more than the lens does. The thickness of the cornea is fixed. The thickness of the lens can change.
5. The *optic nerve* is the nerve that transmits visual stimuli from the eye to the brain.
6. Locate the four *externally attached muscles*. These muscles respond to control eye movement and help focus images. Identify all four muscles.

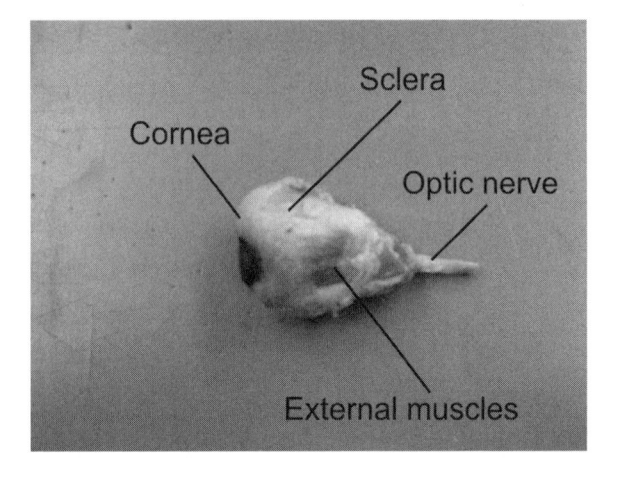

Investigation 17B The Mammalian Eye

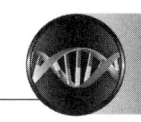

3. Making observations

a. What are the functions of the sclera, cornea, and optic nerve?

b. The human eye has six externally attached muscles instead of only four like the sheep's. Explain how a human's eye might move differently than a sheep's eye.

c. One of your eyes is dominant over the other eye. Form a circle with your thumb and index finger and place your hand in front of you. With both eyes, look at something through the circle. Hold that position and close one eye; then open it. Close the other eye. The eye in which you can still see the object through the circle is your dominant eye. Which of your eyes is dominant?

4. Removing the cornea

1. Use the scissors to carefully cut around the junction between the cornea and the sclera. Cut completely around and remove the cornea.
2. Lay the cornea on the dissecting tray. Identify the lens, iris, and pupil.
3. You will also see a clear jelly-like fluid. This is the vitreous humor. The vitreous humour is attached to the lens.
4. The *lens* is a transparent structure in the eye that, along with the cornea, helps to refract and focus light. A ring of tiny *ciliary muscles*, located along the inner side of the iris, connects the lens to the middle layer of the eye. Ciliary muscles contract to change the curvature of the lens.
5. The *iris* is the pigmented part of the eye and helps to change the size of the pupil.
6. The *pupil* is a hole in the iris. The iris opens or closes to control the amount of light entering the pupil. The pupil gets smaller in bright light and larger in dim light.
7. Remove the lens to look at it. In a living organism, it is completely transparent. To focus on closer objects, the lens gets thicker and more rounded. To focus on faraway objects, it becomes elongated and thinner.

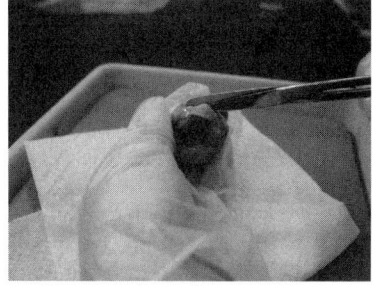

Removing the cornea

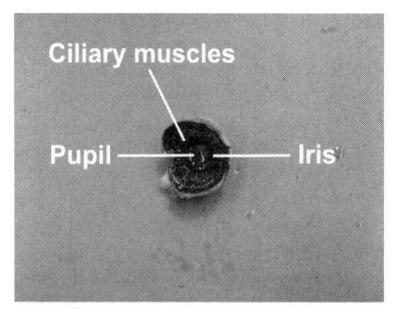

Ciliary muscles
Pupil — Iris

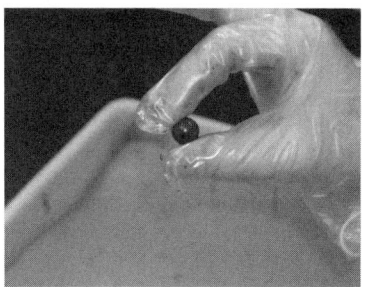

5. Constructing explanations

a. The lens, iris, and pupil are adapted to work together for what purpose?

b. If you enter a very bright room after being in the dark, what would happen to your pupils?

The lens

6. Cutting the eye in half

1. Carefully insert the dissecting scissors at the top of the eye where you removed the cornea.
2. Following a circular pattern around the sclera, rotate the eye while continuing to carefully cut the eye in half.
3. Carefully open and separate the front from the back of the eye without damaging its internal structures.
4. Cutting the sclera in a circular pattern resulted in also cutting the thin reflective middle layer lining of the sheep's eye. This reflective lining is called the *tapetum*. It is not found in the human eye. The tapetum is an adaptation for seeing better in the dark.
5. Observe a wrinkled sac-like structure attached to the optic nerve and connected to the back of the eye. This is the retina. It is the inner most layer of the eye. The <u>retina</u> is a thin layer of cells at the back of the eye of vertebrates. It is the part of the eye that converts light into nerve signals. The point at which the retina tissue connects to the optic nerve is the eye's blind spot.
6. Separate the retina from the back portion of the eye and again observe the colorful reflective layer of the tapetum.
7. Remove the tapetum from the tough, outer layer of the sclera to expose the choroid. The <u>choroid</u> is the layer of the eye lying between the retina and the sclera. The choroid provides oxygen and nourishment to the outer layers of the retina.

The eye cut in half

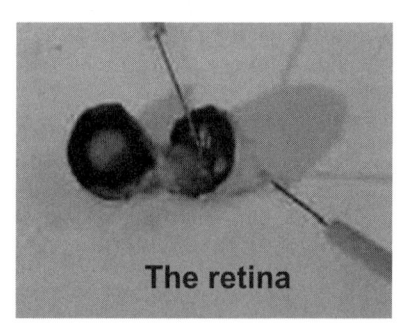

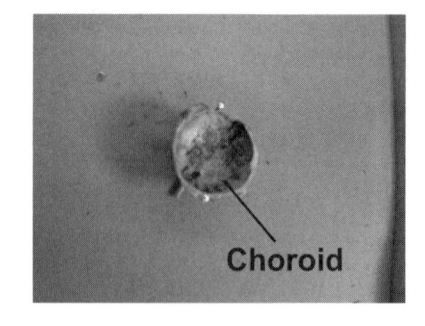

Tapetum

The retina

Choroid

7. Cleaning up

1. Wrap the sheep's eye and its parts in many layers of paper towels.
2. Thoroughly clean all dissecting tools with soap and water.
3. Properly dispose of the paper towel containing the sheep's eye and gloves in the plastic trash bag your teacher brings around.
4. Wash down all lab areas with soap and water.
5. Thoroughly wash your hands with warm water and soap.

Investigation 17B   The Mammalian Eye

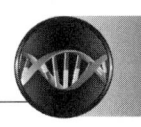

8. Evaluating information

a. Name each structure of the sheep's eye next to its function. Choose three eye structures to classify by tissue type. Add the letter (C) for connective, (M) for muscle, or (N) for nerve next to each structure in Table 1.

Table 1: Structures of the eye and their functions

Eye structure	Function
1.	A. A clear structure that refracts light and can change its curvature
2.	B. Provides oxygen and nourishment to the outer layers of the retina
3.	C. A tiny ring of muscles that change the shape of the lens
4.	D. The pigmented ring of muscles that change the size of the pupil
5.	E. Works with the lens to refract light and helps the eye to focus
6.	F. Move the eye around
7.	G. Transmits signals from the eye to the brain
8.	H. Gives the eye its shape and protects the inner parts
9.	I. A thin layer of cells that convert light into nerve signals

b. Name two differences between the sheep's eye and the human eye.

c. Why does the optic nerve cause a blind spot?

d. To find your blind spot, use the two dots below. Hold one hand over your left eye, and look directly at the left-hand dot. At first, you can see both dots even though you're looking directly at only one. As you slowly move the page closer to your eyes, the right-hand dot disappears! If you move your eye, the dot will reappear, but as long as you focus on the first dot, the second will be invisible. Move even closer and the missing dot reappears. You've found your blind spot!

18A Who's Got the Beat?

How fast will your pulse increase with physical activity?

Unlike most workers, the heart never gets time off. Each hour, an average heart pumps about 75 gallons of blood throughout the body. Even more difficult, during exercise the heart must work overtime. The heart is the pump that causes your blood to circulate throughout your body and to all of your cells. The heart makes sure that the oxygen you breathe, the nutrients from the food you eat, and the water you drink get delivered to your body. Without your heart, your other organs would not be able to function. When you exercise, you increase your body's demand for energy. Therefore, your heart must increase the speed at which the blood delivers the vital nutrients. But just how hard is your heart able to work? In this investigation, you will examine how much your heart rate will increase and how long it takes for your heart to return to its normal rate.

Materials
- Graph paper
- Watch or clock with a second hand
- Metric ruler

1. Planning your investigation

<u>Pulse</u> refers to the vibrations created each time your heart pumps blood through your arteries. Your pulse rate indicates the speed of your heart pumping. You will work with a partner to find your pulse. Follow the steps below to find the pulse in the wrist.

1. Find a large tendon on the underside of your wrist. Using your middle and ring finger from your other hand, locate your pulse just below the base of the thumb and on the outside of the large tendon.
2. While standing, count the number of beats in your pulse while your partner times you for 30 seconds. Then multiply the beats by 2.
3. Practice this step until you can easily find and take your pulse. Repeat at least twice and average your total beats per minute (bpm). This will be your resting pulse rate. Record in Table 1.

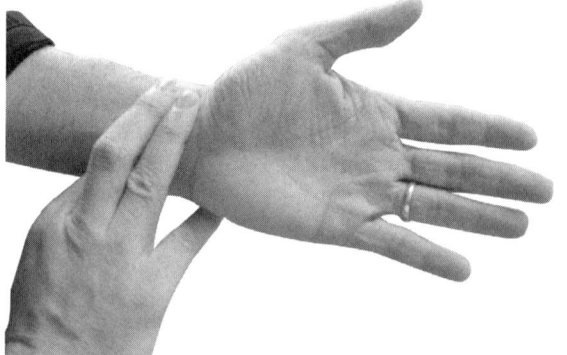

2. Making observations

a. Do you think your pulse is the same whether sitting, standing, or lying down?

b. How much faster do you think your pulse will be if you jump in place for 1 minute? Do you think your partner's pulse will increase by the same amount?

c. Each student in your class will conduct this experiment. What should be the control variables for each student if the class is going to share their data?

Investigation 18A   Who's Got the Beat?

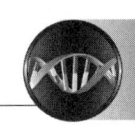

3 Doing the experiment

Safety Note: Caution should be taken with students for whom strenuous activity is risky.

1. Read the next three steps before starting, otherwise you might have to repeat the experiment!
2. Do 50 jumping jacks. Do the jumping jacks in a row without stopping. After you stop, take your pulse for 30 seconds and multiply by two.
3. Continue to take your pulse for the next 5 minutes, waiting 30 seconds in between measurements. This will determine how quickly your heart returns to its resting rate. Record the results after each minute in Table 2.
4. Switch roles and repeat steps 1-3.

Table 1: Rate of each student's pulse when resting

Individual	Beats/minute			
	Trial 1	Trial 2	Trial 3	Average

Table 2: Rate of each student's pulse after exercising

Individual	Beats/minute				
	1 min. after exercising	2 min. after exercising	3 min. after exercising	4 min. after exercising	5 min. after exercising

4 Developing and using models

a. Plot each of your points from Table 2 on a graph. Plot your partner's points in a different color. Use the metric ruler to create a best-fit line.

b. Extend the line to estimate your heart rate after 10 minutes of resting. Using the graph, estimate how long it would take for your heart to return to its resting rate. Your resting rate was calculated in Table 1.

c. Why would it be inaccurate to continue the graph line until it hits 0?

d. What else, besides exercise, might speed up your heart rate?

e. Did you and you partner have the same results? Why might that be?

f. Why do you think your heart rate increases when you exercise?

5. Extending your investigation

a. Your pulse can also be taken on your neck. Would the experiment turn out differently if you used the pulse on your neck instead of your wrist? Design an experiment to find out. Check with your teacher before conducting the experiment.

b. As we get older, does our heart rate change? Design an experiment to compare the average heart rate of individuals at different ages. Will there be a pattern? Check with your teacher before conducting the experiment.

Investigation 18B   The Pressure's On

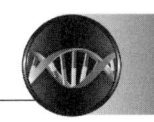

18B The Pressure's On

How much does your blood pressure increase when your heart contracts?

Flexibility is important, especially if you are a blood vessel such as an artery. The average heart pumps 60 to 80 times a minute. With each pump, the heart forces an average of 60 milliliters of blood into attached arteries to be circulated throughout the body to deliver essential oxygen, food, and water. But how much force is necessary with each contraction? Just how much pressure does the heart have to overcome to circulate the 5 liters of blood found in your body? In this investigation, you will find an answer by measuring your blood pressure when your heart is at rest, called diastolic blood pressure, and your blood pressure after your heart just pumped, systolic blood pressure. Then you will compare your results with the average statistics for your age group.

Materials
- Safety goggles
- Sphygmomanometer
- Stethoscope
- Table
- Isopropyl alcohol (50%)
- Cotton balls

Safety Note: Wear goggles when using alcohol to clean the stethoscope ear pieces.

WARNING — This lab contains chemicals that may be harmful if misused. Read cautions on individual containers carefully. Not to be used by children except under adult supervision.

1. Planning your investigation

A sphygmomanometer can measure the amount of pressure required to circulate your blood. When your heart beats, your blood pressure against the arteries increases. This is called _systolic blood pressure_. The heart rests in between beats, so pressure decreases. This is called _diastolic blood pressure_. You will measure both.

1. Sit with your partner at a table. Place your partner's arm on the table with the palm up, and place the cuff of the sphygmomanometer above your partner's elbow. The cuff should be snug but loose enough to fit two fingers through.

2. Dampen a cotton ball with alcohol and clean the stethoscope bell and ear pieces. Place the stethoscope in your ears. Place the round head of the stethoscope on the brachial artery on your partner's arm (right).

3. Make sure the valve is closed on the pump of the sphygmomanometer.

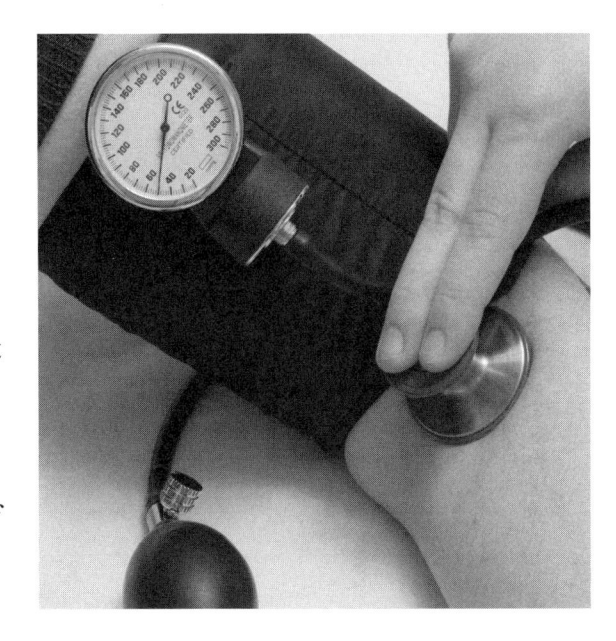

Safety Note: The sphygmomanometer should only be used under adult supervision.

2 Making observations

a. Why might it be important to place your arm in a horizontal position rather than letting it hang at your side?

b. Each student in your class will be tested. What should be the control variables for each student if the class is going to share their data?

c. Make a prediction. Do you think everyone's heart creates the same pressure? Explain why or why not.

3 Doing the experiment

1. Pump enough pressure to bring the needle on the gauge of the sphygmomanometer to 180. Now the pressure from the cuff is greater than the blood pressure so no blood will be able to flow through the artery.
2. Slowly open the valve on the pump, watch the gauge, and listen for a heartbeat. Once the blood pressure is greater than the pressure from the cuff, you will hear the heartbeat. This is the systolic blood pressure, or the blood pressure when it is strongest because the heart has just pumped it. Record this number in Table 1.
3. Continue to listen. Watch the needle on the gauge. Note the number of the needle when you no longer hear the heartbeat. This is the diastolic blood pressure.
4. Record your partner's blood pressure. Clean the ear pieces with alcohol and trade places.

Table 1: Systolic and diastolic blood pressure

Student	Systolic blood pressure (measured in millimeters of mercury)	Diastolic blood pressure (measured in millimeters of mercury)
You		
Your partner		

Table 2: Average blood pressure of individuals, by age

Age (years)	Average systolic pressure (measured in millimeters of mercury)	Average diastolic pressure (measured in millimeters of mercury)
Birth	70	45
5	?	?
10	105	70
15	117	77
20	120	79
25	121	80
30	122	81
35	123	82
40	125	83
45	127	84
50	129	85
55	131	86
60	134	87

Investigation 18B *The Pressure's On*

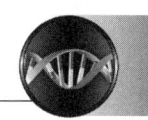

4. Developing and using models

a. Plot each of the points from Table 2 onto a graph. Plot diastolic and systolic blood pressures in different colors. Use the metric ruler to create a best-fit line.

b. What would be the systolic and diastolic blood pressure for the average 5 year old?

c. Why might the blood pressure be so low for an infant?

d. How does blood pressure change with age? Explain why this might be.

e. What can an individual do to increase his or her blood pressure?

f. What can an individual do to decrease his or her blood pressure?

g. Arteries are blood vessels that carry blood away from the heart. Veins are blood vessels that carry blood to the heart. Would you expect the blood pressure to be greater in your arteries or in your veins? Explain why.

5. Extending your investigation

Does your blood pressure change with the time of day? How could you investigate this question? Design an experiment to find out. Check with your teacher before conducting the experiment.

19A Levers

How does a lever work?

How could you lift up a car—or even an elephant—all by yourself? One way is with a lever. The lever is an example of a simple machine.

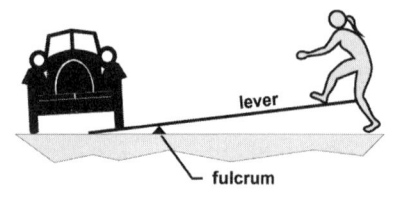

Materials
- Lever kit (with screw, bolt, and yellow strings)
- Physics stand
- Weights (set of 8)

Safety Note: Only use the third hole from the bottom of the physics stand to attach the lever. Be careful when handling heavy weights. They can injure fingers or toes if dropped.

1 Planning your investigation

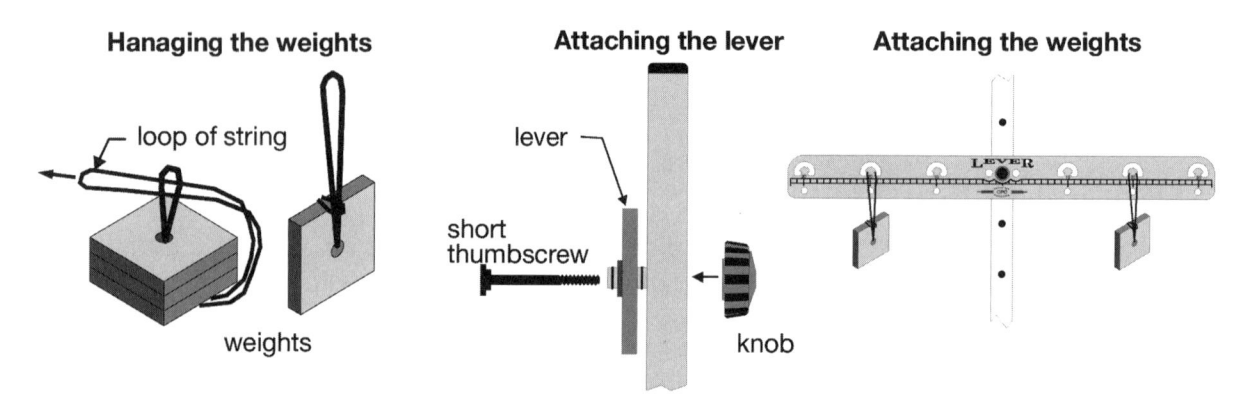

1. Place the lever on the third hole from the bottom. Use loops of string to make hangers for the weights. You can put more than one weight on a single string.

2. The weights can be hung from the lever by hooking the string over the center peg in the holes. Make sure that the string is all the way around the peg!

2 Levers in equilibrium

a. The lever is in equilibrium when all the weights on one side balance all the weights on the other side. Hang the weights as shown below. Does the lever balance?

b. What variables can be changed to balance a lever?

c. On the diagram below label the fulcrum, the input arm, and the output arm.

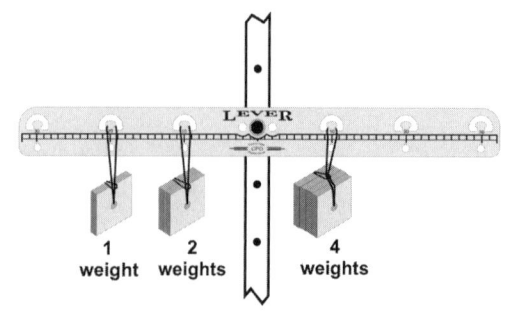

112

Investigation 19A   Levers

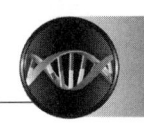

3. Trying different combinations to balance the lever

Make different combinations of weights and positions that balance. Use the chart below to write down the numbers of weights you put in each position. If you want to conduct more than four trials, write your results on a separate sheet of paper.

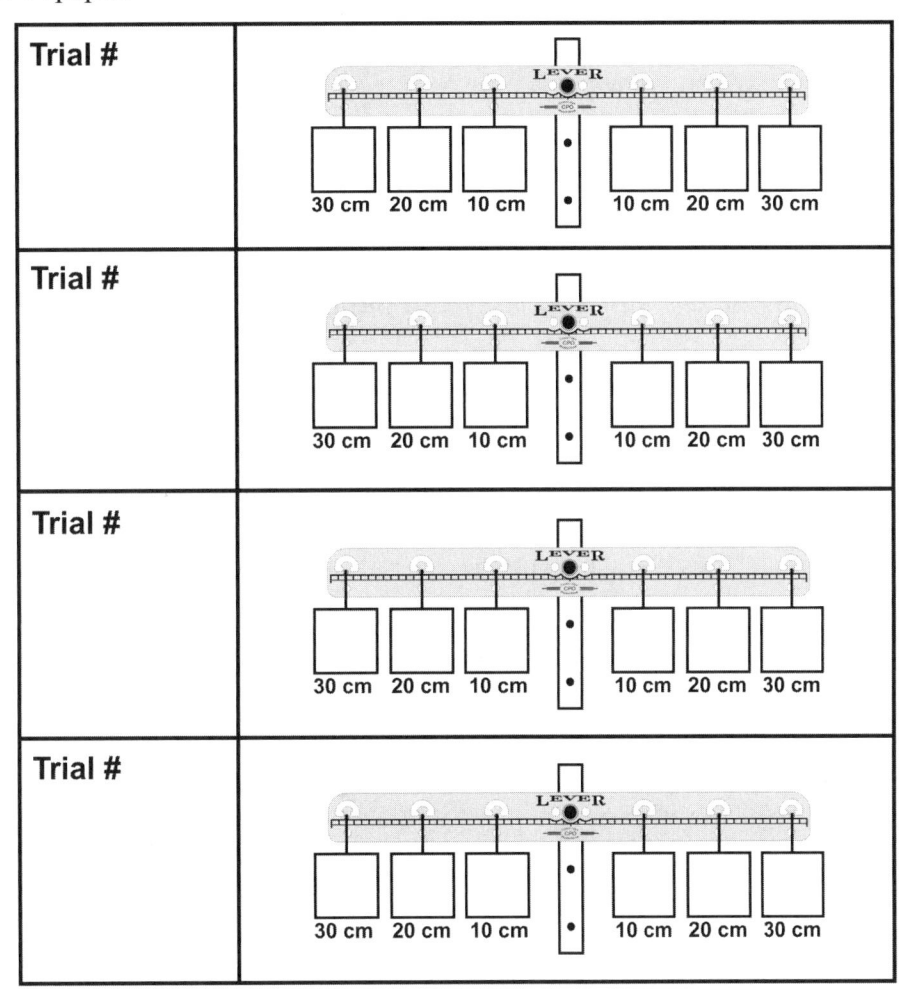

Trial #	
Trial #	
Trial #	
Trial #	

4. Determine the mathematical rule for equilibrium

Using the data in the chart above, determine a mathematical rule for levers in equilibrium. Think about the variables in the experiment: *input force*, *output force*, length of input arm, and length of output arm. First, make some calculations, then write your rule as an equation.

5. Developing models

a. Draw a lever and label these parts: fulcrum, input arm, output arm, input force, and output force.

b. There are two ratios that can be used to determine mechanical advantage in levers. What are the two equations? What is the relationship between the two equations?

c. What could you do to the input side of a lever to increase the amount of output force? (HINT: There are two different things you could do.)

113

19B Levers and the Human Body

How does the human arm work?

Arms, legs, fingers, toes, the jaw, even the head and neck work like levers. Contracting and extending muscles provide the force to move our levers. Our joints are the fulcrums around which these levers pivot and move. Our bones are the levers themselves. In this investigation, you will look at the human arm and examine how it works like a lever.

Materials

- Lever kit (with screw, bolt, and yellow strings)
- Physics stand
- 4 Spring scales (2.5, 5, 10, and 20 N)
- Weights (set of 8)

Safety Note: Only use the third hole from the bottom of the physics stand to attach the lever. Be careful when handling heavy weights. They can injure fingers or toes if dropped.

1. Planning your investigation

Let your left arm hang down by your side. Place your right hand into the inner part of your left elbow. Slowly lift your left forearm (palm up) until it is level with the floor. Feel that tissue in your elbow tightening up with your right hand? That is the connective tissue that joins your biceps muscle to the bones of your forearm, the radius and ulna. We can use the physics stand and the lever to make a model of the human arm and measure the forces involved when we lift something.

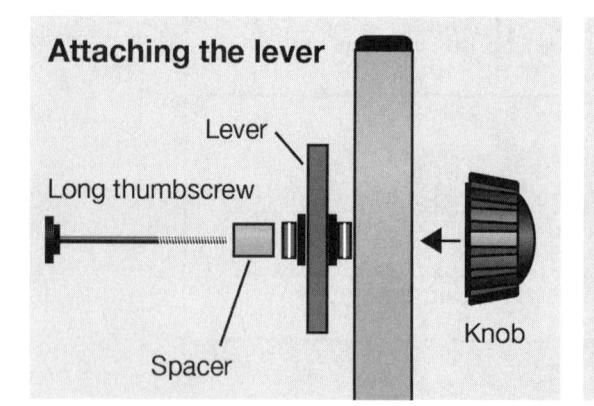

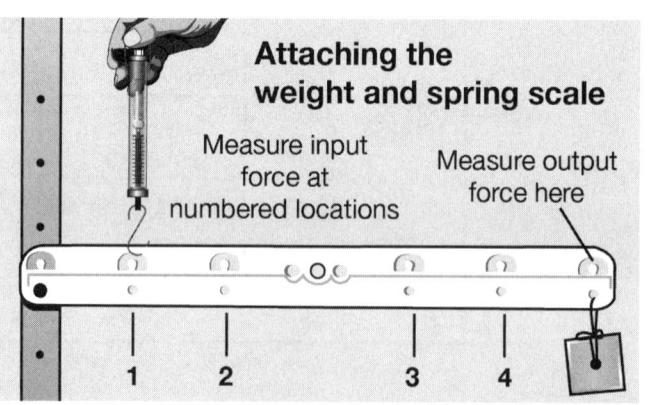

1. Attach the lever to the stand on the third hole from the bottom, but this time use the hole on the left-most side of the lever. Use one of the long thumbscrews. Do not tighten the knob all the way. Leave a little room so the lever can still pivot.

2. Use a loop of string to hang one weight on the right-most hole of the lever. Use the green spring scale to measure the downward force in newtons just above the weight. This is the output force for the experiment, and this variable will not change. Record the output force in column one of Table 1.

3. Measure the output distance, in centimeters, of the lever. The output distance is the distance from the fulcrum to the output force. The output distance will be the same for each trial. Record the output distance in column two of Table 1.

4. Study the diagram above and then answer the questions in Part 2.

Investigation 19B *Levers and the Human Body*

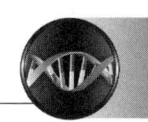

Table 1: Input and output data

Trial	Output force (N)	Output distance (cm)	Input force (N)	Input distance (cm)	Mechanical advantage
1					
2					
3					
4					

2. Developing a model

a. Label the fulcrum, output force, output distance, input force, and input distance on the diagram in Part 1.

b. The lever you are testing in this experiment is a model of the human arm. Which part of the lever represents the location of the hand?

c. Predict the position on the diagram where the mechanical advantage of the arm would be greatest and least.

3. Making observations

1. Measure the input force at position 1. The input force is the force required to lift the lever up and keep it horizontal. Record your results in the first row, third column of Table 1.

2. Measure the input distance for position 1. Record your results in the first row, fourth column of Table 1.

3. Measure the input force and input distance for the remaining positions, and record your results in Table 1.

4. Calculate the mechanical advantage (MA) for each trial using the formula below. Record the value for each trial in the last column of Table 1.

$$MA = \frac{\text{output force (N)}}{\text{input force (N)}}$$

4. Constructing explanations

a. At which position is the mechanical advantage greatest? Least?

b. Which trial represents the position of the fulcrum, input force, and output force of the human arm?

c. Is the human arm adapted to have the greatest mechanical advantage or the greatest output distance? Explain why the human arm is adapted in this way.

d. Draw a diagram of the human arm lifting a weight. Label the fulcrum, input and output force, and input and output distance.

20A Color Vision

How do our eyes see color?

Most mammals have limited color vision. Humans are one exception. Special cells located on the retina work together with other structures in the eye to enable humans to see color. The color you "see" depends on how much energy is received by three different types of cells. In this investigation, you will use a white light source and color filters to discover what happens when you mix the primary colors of light. How does this help explain how your eyes perceive color?

Materials

- Light and Color kit
- (6) AA batteries
- Room that may be darkened
- TV monitor or computer screen
- Metric measuring tape

1. Sources of light

1. There are two different sources of light for vision. Light can be *produced by* objects or *reflected by* (bounce off) objects. Look for clothing in your class that is a solid color.

a. How does light help you see colors of clothing?

b. Does the light reaching your eye from the clothes come from the clothes, or does the light come from somewhere else?

c. Your teacher will turn off all the lights and shade the windows so it is darker in the room. How did the color of the clothing change?

d. Your teacher will now turn on a monitor in a dark room. How is viewing the monitor different from viewing clothing in a darkened room?

e. Your observations provide evidence for the two different ways light reaches the human eye. In complete darkness, can you see clothing?

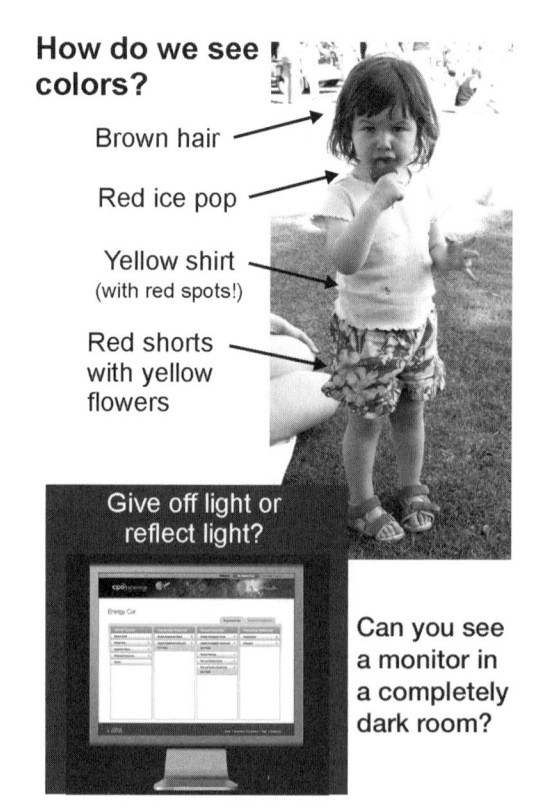

How do we see colors?

- Brown hair
- Red ice pop
- Yellow shirt (with red spots!)
- Red shorts with yellow flowers

Give off light or reflect light?

Can you see a monitor in a completely dark room?

Investigation 20A   Color Vision

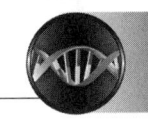

2. Mixing primary colors of light

1. Connect two holders by sliding one rail into the slot of another. Lay the third holder on its side to make a pyramid.

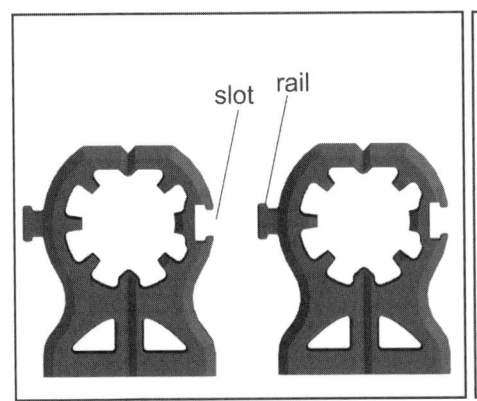

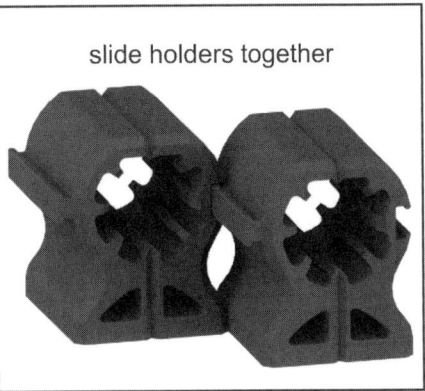

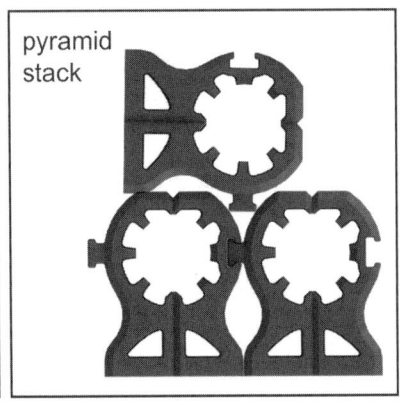

2. Place three colored filter caps on three different flashlights, and check to be sure they have AA batteries before sliding each into a holder. You can turn the lights on or off with buttons on their ends.

3. Use the box that the Light & Color kit comes in as a white screen. If needed, fold the top of the box over to shade the area from any overhead lights. Place the flashlight pyramid about 40 cm from the white box. Turn on all the flashlights.

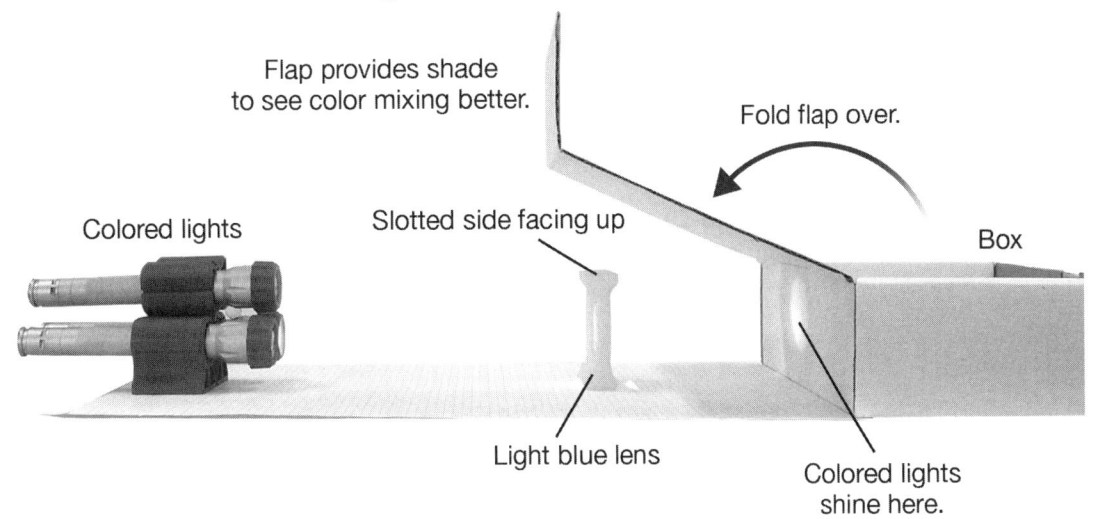

4. Locate the lens and set it just in front of the flashlights so they shine directly through the lens. Place the lens so the slotted side with openings faces up.

5. Slowly move the lens away from the lights and toward the box until you see three overlapping circles of color in focus on the box.

3. Making observations

a. What color do you see where the red and green lights overlap?

b. What color do you see where the green and blue lights overlap?

c. What color do you see where the blue and red lights overlap?

117

d. What color is produced when all three colors of light overlap in the middle?

Table 1: Mixing primary colors of light

LED color combination	Color you see
Red + Green	
Red + Blue	
Blue + Green	
Red + Green + Blue	

4. Developing a model

Color is how humans perceive the energy of visible light. Red has the lowest energy and violet has the highest. White light consists of a range of colors, sometimes abbreviated as ROYGBIV.

a. Turn off the red and green flashlights and remove the blue cap. What color light do you see through the lens?

b. What evidence do you have that white light contains different colors?

c. Where do you think the energy comes from for humans to see color?

White light travels through the cornea, enters the pupil in your eye, and is focused by the lens on specific tissues in the retina. The retina consists of specialized light sensitive rod cells and cone cells. Rod cells allow humans to see better in dim or low level light. Cone cells are sensitive to red, blue, and green light.

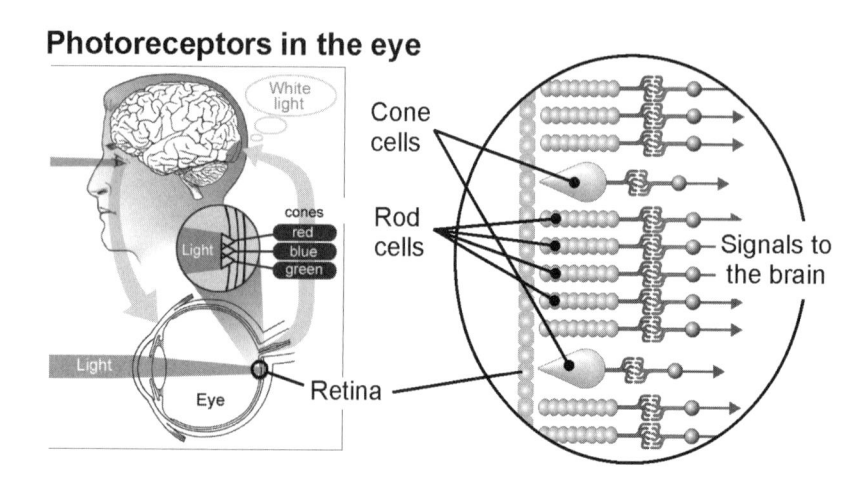

Photoreceptors in the eye

a. Think back to part one when you observed clothing with the overhead lights in the classroom turned off. Which cells became active for under low light conditions?

b. Why do you think nocturnal mammals evolved to have lots of rod cells but few or no cone cells at all?

c. There are no specialized cells in humans to detect the individual energy of yellow, magenta (pink), or cyan (aqua). Argue from your evidence why these cells are not needed for human vision.

Investigation 20A Color Vision

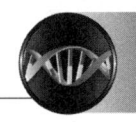

5. Extending your investigation

a. Research online to explain color vision in nocturnal mammals. Consider the color the eyes "glow" at night. How many cone cell types do they have? Use the following terms in short presentation to the class: cone cells, nocturnal, retina, and rod cells.

b. Sunlight contains ultraviolet light, also known as ultraviolet radiation. This is invisible to the human eye. Research online about what exposure to UV rays can do to human epidermis. Prepare a short presentation for the class. Use the terms: energy, epidermis, mutate, and radiation in your report.

20B Light and Vision

How does the lens in the human eye help us see an image?

A lens is a transparent disc designed to refract (bend) light in a specific way. Many devices use lenses, such as glasses, cameras, microscopes, and telescopes. A typical lens in an object is a fixed shape and usually made of glass. In the human eye, the lens is a more flexible structure. Several parts in the human eye work together with the lens and the nervous system to help you see images. In this investigation you will model how the human eye forms an image.

Materials
- Light and Color kit
- (2) AA batteries
- Metric measuring tape
- White wall or white paper & tape
- Magnifying lens (6x)
- Graph paper (5 mm^2)

1. Projecting an image with a lens

Recall that animals have adapted different cells, tissues, organs, and systems to work together for vision. In a perfect eye, all the light collected from every point on an object is refracted (bent) so it comes together to construct duplicate image on the back of the eye. In this part of the investigation, you will model how light can enter a lens like the eye and can project an image onto a surface.

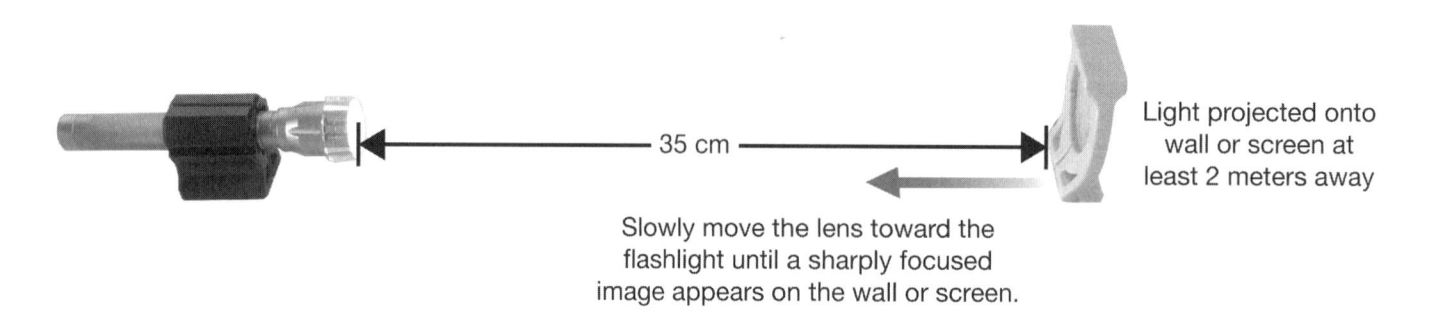

35 cm

Light projected onto wall or screen at least 2 meters away

Slowly move the lens toward the flashlight until a sharply focused image appears on the wall or screen.

1. Put one of the flashlights in a holder and place it on a table. Put the clear filter with the letter "F" on the flashlight.
2. Point it toward a white or light-colored wall, at least 2 meters away. If a white wall is not available, tape a piece of white paper to a flat surface to be your screen.
3. Set the lens with the light blue frame about 35 cm away from the front of the flashlight.
4. Turn on the flashlight and position it so it shines through the lens onto the wall or paper.
5. Slowly move this lens toward the light until you see a sharp image of the "F" on the wall or paper. Have one group member check the projected image closely while the lens is slowly moved. Discuss with your team the exact place the lens needs to be to make the letter come into focus.

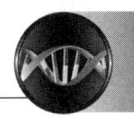

Investigation 20B   Light and Vision

2. Developing your model

a. Our simple model of a human eye has several limitations. Through what covering and opening in the eye does light travel through first? Both are missing from our model.

b. What structure in the human eye might correspond to the frame around the lens that enables the human lens to change shape to focus?

c. Describe the image of the letter produced on the wall. How does it compare to the letter's position on the flashlight cap?

d. The purpose of the lens in a human eye is to focus light on specialized layers of tissue that contains light sensitive cells. Predict which part of your model you think represents this structure? What is its name?

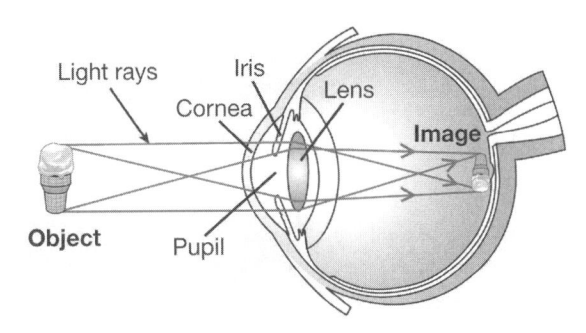

3. Focusing light rays

The lens in our model and in the human eye are both called converging lenses because light rays entering one side of the lens are bent so that they all "converge," or cross in the same place on the other side of the lens to create an image. The focal point is a place where all the rays meet. Distances to and from focal points are how we distinguish between types and magnification of lenses. Some lenses are also double convex lenses because both sides of the lens have the same exact curved surface.

1. Examine the light blue lens and compare it to the hand lens provided. Hold each lens between your thumb and forefinger to compare the curvature of two lenses. Are they the same? Record your observations in Table 1.

2. Lay a piece of centimeter graph paper on your table.

3. Place the light blue lens flat on the paper. Count how many squares you see across the widest part through the center, or its diameter. Record it in Table 1. Repeat this step using the hand lens. Record.

4. Now position your head at the top of the ruler and the blue lens near the bottom. While looking through the lens, keep your eye in the same spot at the top of the ruler and raise the lens slowly from the paper. When the grid comes into focus, count the number of squares you see. Record the distance from the graph paper at the center of the lens.

5. Repeat step 4 for the hand lens. Be sure to raise it to the same height on the ruler.

121

6. Calculate the magnification for each lens by dividing the total number of blocks seen on the paper by the total number seen through the raised lens. Record in Table 1.

Magnification of 4.5

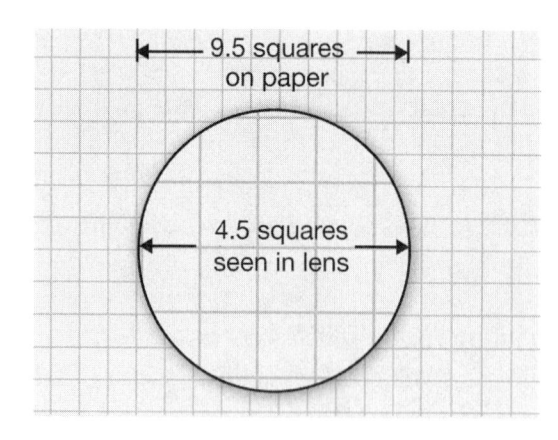

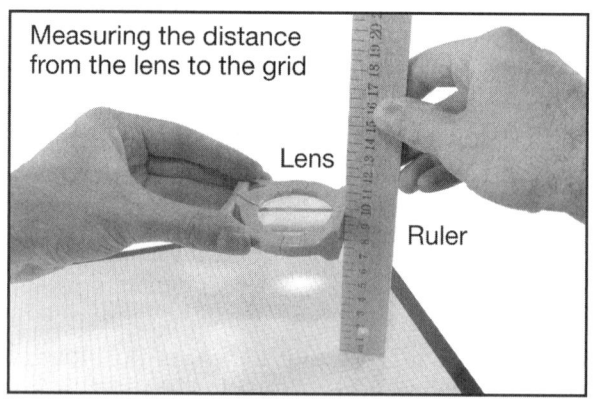

Table 1: Lens diameter and magnification

	Description of lens shape	Diameter with lens on paper (no. of squares)	Diameter with lens raised (no. of squares)	Distance raised from the paper (cm)	Magnification (col 2 ÷ col 3)
Light blue lens					
Hand lens					

4. Constructing explanations

a. Which lens magnified better, the light blue lens or hand lens?

b. The hand lens has a stated magnification of 6x. How close did your estimate come to this value? Why might the values be different?

c. Increasing the curvature of a lens can affect the final image. Explain.

5. The retina and correcting bad vision

Where the light rays converge is important to human vision. If you predicted in Part 3 that the retina was the flat area where the image formed on the wall or paper, you predicted correctly! If the lens in your eye doesn't focus the image on the retina, you might have trouble seeing it.

1. Consider what might happen to a person's vision if the lens can't focus light rays on the retina. Move flashlight with the clear cover and the lens back to its position 35 cm away and facing the wall or paper. You should see the letter in focus.
2. Move the lens 4-5 cm towards the flashlight. The letter should no longer be visible.
3. Place the hand lens in between the flashlight and the lens. You may need to make minor adjustments to both lenses, but you should see the letter again.

Investigation 20B *Light and Vision*

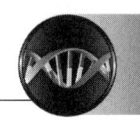

   a. Did you see any difference between using one lens and two lenses? Explain.

   b. How can placing another lens in front of an eye change vision?

   c. Can you think of two different ways people correct bad vision?

6 Extending your investigation

a. Use the flashlight, lens, and screen to experiment with different distances between the three objects to see how many places images can be focused and whether the image is the same. Record your procedure, make a data table, and report your findings.

b. The making of lenses for better vision began a long time ago. Choose a topic to explore and report back to your teacher or class.

- Anton Von Leeuwenhoek was a Dutch lens maker. Learn more about his life and contributions to life science.

- New technology has helped us diagnose and treat various vision problems. Explore a technology and explain how people have benefitted. What are the risks involved with using the new technology?

LAB SKILLS

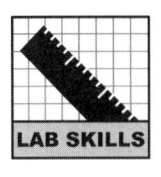

Safety Skills 126
Writing a Lab Report 132
Measuring Length 135
Measuring Temperature 139
Calculating Volume 141
Measuring Volume 145
Measuring Mass with a Triple Beam Balance 146
Using a Compound Microscope 149
Recording Observations in the Lab 154

EXPLORE further

Measuring Biodiversity 158
Innovation and Recycling 161

Safety Skills

What can I do to protect myself and others in the lab?

Science equipment and supplies are fun to use. However, these materials must always be used with care. Here you will learn how to be safe in a science lab.

Materials
- Poster board
- Felt-tip markers

1. Follow these basic safety guidelines

Your teacher will divide the class into groups. Each group should create a poster-sized display of one of the following guidelines. Hang the posters in the lab. Review these safety guidelines before each Investigation.

1. Listen to all teacher instructions before, during, and after investigations.

2. Prepare for each investigation or activity.
   a. **Sign** the Science Safety Student Responsibility Agreement.
   b. **Read** each activity or investigation carefully.
   c. **Identify** the investigation purpose.
   d. **Work** ONLY on activities approved by your teacher.
   e. **Follow** all oral and written safety instructions.
   f. **Know** the location of Emergency Safety Equipment such as fire extinguisher, eye and face wash station, safety shower, and first aid kit.

3. Dress for laboratory work.
   a. **Wear** protective equipment such as chemical splash goggles, laboratory aprons, and protective gloves as needed.
   b. **Roll** long sleeves above the wrist.
   c. **Tie** back long hair.
   d. **Remove** dangling jewelry and any loose or bulky outer layers of clothing.
   e. **Wear** shoes that enclose the feet. (no flip flops, sandals or open-toe shoes).

4. Prevent unsafe situations.
   a. **Be aware** of classmates and their safety.
   b. **Do not** touch, taste or smell any substance without teacher instructions.
   c. **Never** work alone in the laboratory.
   d. **Don't enter** science or chemical storage or preparatory areas without a teacher.
   e. **Keep** your work area clean and uncluttered.

5. **Dispose** of used or unused materials according to teacher directions.

6. Return equipment to the proper location.

7. Wash hands with soap and warm water for 20 seconds after experimenting.

8. Act wisely in an emergency.

   a. **Notify** your teacher of any accident immediately.

   b. **Follow** emergency procedures in event of accident.

2. Know what to do when...

1. working with glassware.

   a. **Don't use** glassware that is chipped or cracked.

   b. **Use special care** to prevent breakage and cuts or scratches.

2. working with heat.

   a. **Wear** heat-resistant gloves at all times.

   b. **Do not** touch hot items with bare hands. Use heat-resistant gloves, pads, or tongs.

   c. **Heat** water only in open containers of heat-resistant glass.

   d. **Watch** all burners, hot plates or open flames.

   e. **Warn** others if they come close to your hot items or liquids.

3. working with electricity.

   a. **Keep** electric cords away from water.

   b. **Don't** use frayed cords or plugs in outlets.

4. finished experimenting.

   a. **Return** clean materials to their proper locations.

   b. **Dispose** of all used solids and liquids according to teacher instructions. Do not put items in trash or wash down sink without permission.

   c. **Wash** your hands with soap and water for 20 seconds.

5. you have safety concerns. Tell your teacher, or get help immediately if:

   a. **You have** trouble using your equipment.

   b. **You do not** understand the instructions for the investigation.

   c. **You injure** yourself, or see someone injured.

   d. **You see** or smell something burning.

   e. **You smell** chemical or gas fumes.

▲ Safety quiz

1. Draw a diagram of your science lab in the space below. Include in your diagram the following items. Include notes that explain how to use these important safety items.

   - Exit/entrance ways
   - Fire extinguisher(s)
   - Fire blanket
   - Eye wash and shower
   - First aid kit
   - Location of eye goggles and lab aprons
   - Sink
   - Trash cans
   - Location of special safety instructions

2. How many fire extinguishers are in your science lab? Explain how to use them.

3. List the steps that your teacher and your class would take to safely exit the science lab and the building in case of a fire or other emergency.

4. Before beginning certain Investigations, why should you first put on protective goggles and clothing?

5. Why is teamwork important when you are working in a science lab?

6. Why should you clean up after every Investigation?

7. List at least three things you should you do if you sense danger or see an emergency in your classroom or lab.

8. Six lab situations are described below. What would you do in each situation?
    a. You accidentally knock over a glass container and it breaks on the floor.

    b. You accidentally spill a large amount of water on the floor.

129

c. You suddenly you begin to smell a "chemical" odor that gives you a headache.

d. You hear the fire alarm while you are working in the lab. You are wearing your goggles and lab apron.

e. While your lab partner has her lab goggles off, she gets some liquid from the experiment in her eye.

f. A fire starts in the lab.

Safety in the science lab is everyone's responsibility!

4. Safety contract

Keep this contract in your notebook at all times.

By signing it, you agree to follow all the steps necessary to be safe in your science class and lab.

I, _____, (Your name)

- ☑ Have learned about the use and location of the following:
  - Aprons, gloves
  - Eye protection
  - Eyewash fountain
  - Fire extinguisher and fire blanket
  - First aid kit
  - Heat sources (burners, hot plate, etc.) and how to use them safely
  - Waste-disposal containers for glass, chemicals, matches, paper, and wood
- ☑ Understand the safety information presented.
- ☑ Will ask questions when I do not understand safety instructions.
- ☑ Pledge to follow all of the safety guidelines that are presented on the Safety Skills sheet at all times.
- ☑ Pledge to follow all of the safety guidelines that are presented on Investigation sheets.
- ☑ Will always follow the safety instructions that my teacher provides.

Additionally, I pledge to be careful about my own safety and to help others be safe. I understand that I am responsible for helping to create a safe environment in the classroom and lab.

Signed and dated,

_____

Parent's or Guardian's statement:

I have read the Safety Skills sheet and give my consent for the student who has signed the preceding statement to engage in laboratory activities using a variety of equipment and materials, including those described. I pledge my cooperation in urging that she or he observe the safety regulations prescribed.

_____   _____

Signature of Parent or Guardian   Date

Writing a Lab Report

How do you share the results of an experiment?

A lab report is like a story about an experiment. The details in the story help others learn from what you did. A good lab report makes it possible for someone else to repeat your experiment. If their results and conclusions are similar to yours, you have support for your ideas. Through this process we come to understand more about how the world works.

1. The parts of a lab report

A lab report follows the steps of the scientific method. Use the checklist below to create your own lab reports.

- ☐ **Title:** The title makes it easy for readers to quickly identify the topic of your experiment.

- ☐ **Research question:** The research question tells the reader exactly what you want to find out through your experiment.

- ☐ **Introduction:** This paragraph describes what you already know about the topic and shows how this information relates to your experiment.

- ☐ **Hypothesis:** The hypothesis states the prediction you plan to test in your experiment.

- ☐ **Materials:** List all the materials you need to do the experiment.

- ☐ **Procedure:** Describe the steps involved in your experiment. Make sure that you provide enough detail so readers can repeat what you did. You may want to provide sketches of the lab setup. Be sure to name the experimental variable and tell which variables you controlled.

- ☐ **Data/Observations:** This is where you record what happened, using descriptive words, data tables, and graphs.

- ☐ **Analysis:** In this section, describe your data in words. Here's a good way to start: *My data shows that...*

- ☐ **Conclusion:** This paragraph states whether your hypothesis was correct or incorrect. It may suggest a new research question or a new hypothesis.

2. A sample lab report

Use the sample lab report on the next two pages as a guide for writing your own lab reports. Remember that you are telling a story about something you did so that others can repeat your experiment.

Name: Jon G. **Date:** October 20, 2016

Title: The effect of temperature on tomato seedling growth

Research question: Will tomato seedlings grow faster in a warmer environment?

Introduction:

I planted sunflower seeds in my garden on May 1 last year. We had a heat wave the third week of May with temperatures averaging 5°C higher than normal. The seedlings seemed to grow like crazy that week. Next spring, I want to plant tomatoes in my garden. Since tomatoes have to be started indoors in our climate, I am going to test the growth rate of tomato seeds at 20°C and 25°C.

Hypothesis: The seedlings kept at 25°C will grow faster than those kept at 20°C.

Materials:

5-pound bag "starter mix" potting soil	1 packet tomato seeds (at least 50 seeds)
tap water	2 "grow-lights"
5 gallon bucket for mixing soil	plant mister
50 10-centimeter diameter plastic pots	graduated cylinder to measure 20 mL water

Procedure:

1. I mixed a 5-pound bag of "starter mix" potting soil with enough water to make it evenly moist, but not soggy.
2. I filled 50 10-centimeter diameter plastic pots with the moist potting soil. The soil has time release plant food in it.
3. I placed one tomato seed into each pot at a depth of about 3 millimeters.
4. I placed all pots under a "grow-light" in a spare room in our house. I kept the shades closed so that the plants would all receive the same amount of light. The thermostat in this room is set at 25°C.
5. The soil in each pot received five sprays with a plant mister (20 mL) each day to keep it moist.
6. After the seeds germinated, I chose 30 plants that were the same height—7 millimeters tall. Each of the chosen plants had just two leaves.
7. I divided these plants into two groups. The fifteen plants in group one were left in the spare room, where the temperature remained 25°C.
8. I brought the fifteen plants in group 2 down to our basement. The temperature there is about 5°C cooler than the rooms upstairs. It remains at about 20°C (plus or minus 0.5°C). I set up a second "grow light" just like the first one. There are no windows in the basement so the plants didn't receive any additional light.
9. I continued to give each plant 20 milliliters of water each day for 28 days.
10. I measured the height of all plants on day 7, day 14, day 21, and day 28.

Data/Observations:

Table 1: Average seedling height

Elapsed time (days)	Group 1 (25°C) Average seedling height (mm)	Group 2 (20°C) Average seedling height (mm)
0 (start)	7	7
7	9	7
14	19	12
21	40	23
28	50	29

Seedling growth at different temperatures

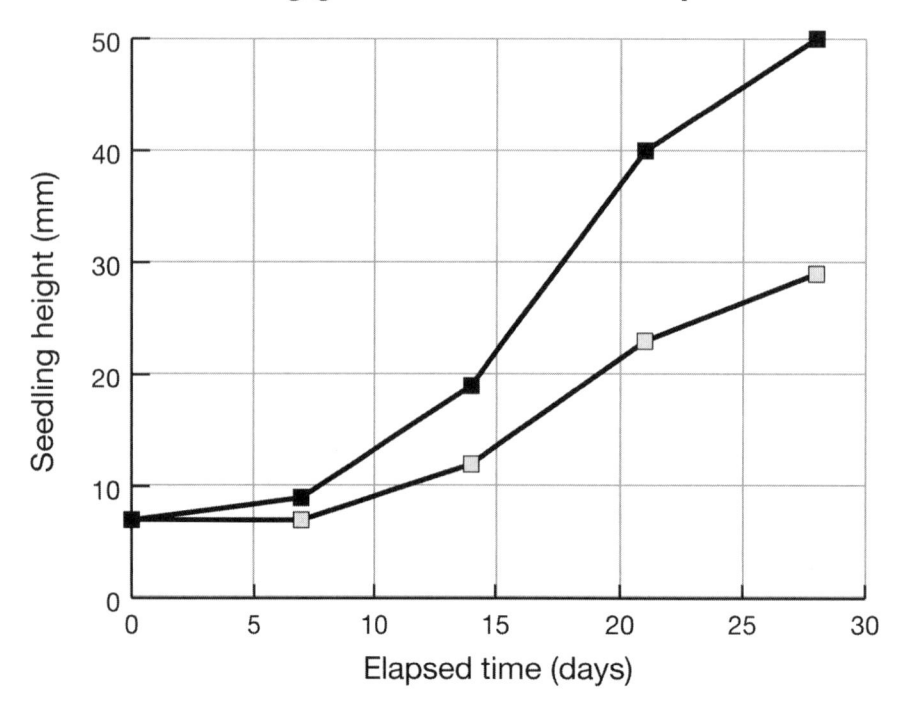

Analysis:

My data shows that the seedlings in the 25°C environment grew faster than those in the 20°C environment. By the 28th day, the seedlings in the warmer room had grown 43 millimeters, while those in the colder room had grown only 22 millimeters.

Conclusion:

The seedlings in the warmer environment grew almost twice as fast! Next, I would like to test the growth rate of tomato seedlings at 30°C and 35°C. I would like to know how warm is too warm. This would help me determine the best temperature for starting tomato seeds.

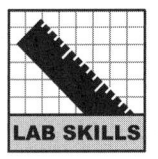

Measuring Length

How do you find the length of an object?

Size matters! When you describe the length of an object or the distance between two objects, you are describing something very important about the object. Is it as small as a bacteria (2 micrometers)? Is it a light year away (9.46×10^{15} meters)? By using the metric system you can quickly see the difference in size between objects.

Materials

- Metric ruler
- Pencil
- Paper
- Small objects
- Calculator

1. Reading the meter scale correctly

Look at the ruler in the picture above. Each short line on the top of the ruler represents one millimeter. Longer lines stand for 5-millimeter and 10-millimeter intervals. When the object you are measuring falls between the lines, read the number to the nearest 0.5 millimeters. Practice measuring several objects with your own metric ruler. Compare your results with a lab partner.

2. Stop and think

a. You may have seen a ruler like the one above marked in centimeter units. How many millimeters are in one centimeter?

b. Notice that the ruler also has markings for measuring with the English system. Give an example of when it would be better to measure with the English system than the metric system. Give a different example of when it would be better to use the metric system.

3. Example 1: Measuring objects correctly

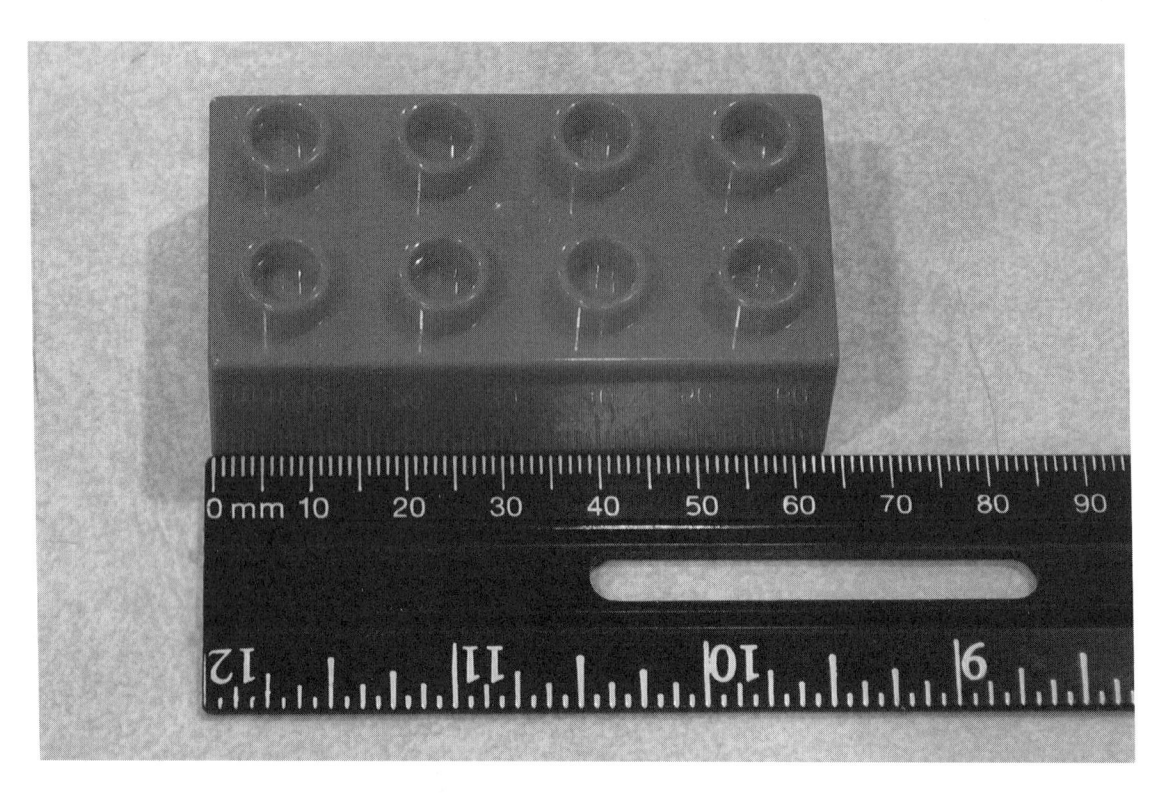

Look at the picture above. How long is the building block?

1. Report the length of the building block to the nearest 0.5 millimeters.
2. Convert your answer to centimeters.
3. Convert your answer to meters.

4. Example 2: Measuring objects correctly

Look at the picture above. How long is the pencil?

1. Report the length of the pencil to the nearest 0.5 millimeters.
2. Challenge: How many building blocks in example 1 will it take to equal the length of the pencil?
3. Challenge: Convert the length of the pencil to inches by dividing your answer by 25.4 millimeters per inch.

5. Example 3: Measuring objects correctly

Look at the picture above. How long is the domino?

1. Report the length of the domino to the nearest 0.5 millimeters.
2. Challenge: How many dominoes will fit end to end on the 30 cm ruler?

6. Practice converting units for length

By completing the examples above, you show that you are familiar with some of the prefixes used in the metric system, like milli- and centi-. The table below gives other prefixes you may be less familiar with. Try converting the length of the domino from millimeters into all the other units given in the table.

Refer to the multiplication factor this way:

- 1 kilometer equals 1,000 meters.
- 1,000 millimeters equals 1 meter.

1. How many millimeters are in a kilometer?

   > 1,000 millimeters per meter × 1,000 meters per kilometer = 1,000,000 millimeters per kilometer

2. Fill in the table with your multiplication factor by converting millimeters to the unit given. The first one is done for you.

   > 1,000 millimeters per meter × 10^{-12} meters per picometer = 10^{-9} millimeters per picometer

3. Divide the domino's length in millimeters by the number in your multiplication factor column. This is the answer you will put in the last column.

Prefix	Symbol	Multiplication factor	Scientific notation in meters	Your multiplication factor	Your domino length in:
pico-	p	0.000000000001	10^{-12}	10^{-9}	pm
nano-	n	0.000000001	10^{-9}		nm
micro-	μ	0.000001	10^{-6}		μm
milli-	m	0.001	10^{-3}		mm
centi-	c	0.01	10^{-2}		cm
deci-	d	0.1	10^{-1}		dm
deka-	da	10	10^{1}		dam
hecto-	h	100	10^{2}		hm
kilo-	k	1,000	10^{3}		km

Measuring Temperature

How do you find the temperature of a substance?

There are many different kinds of thermometers used to measure temperature. Can you think of some you have at home? In your classroom you will use a glass immersion thermometer to find the temperature of a liquid. The thermometer contains alcohol with a red dye in it so you can see the alcohol level inside the thermometer. The alcohol level changes depending on the surrounding tube. You will practice reading the scale on the thermometer and report your readings in degrees Celsius.

Materials
- Alcohol immersion thermometer
- Beakers
- Water at different temperatures
- Ice

Safety Note: Glass thermometers are breakable. Handle them carefully. Overheating the thermometer can cause the alcohol to separate and give incorrect readings. Glass thermometers should be stored horizontally or vertically (never upside down) to prevent alcohol from separating.

1. Reading the temperature scale correctly

Look at the picture at right. See the close-up of the line inside the thermometer on the scale. The tens scale numbers are given. The ones scale appears as lines. Each short line equals 1 degree Celsius. Practice reading the scale from the bottom to the top. One short line above 20°C is read as 21°C. When the level of the alcohol is between two short lines on the scale, report the number to the nearest 0.5°C.

2. Stop and think

a. What number does the long line between 20°C and 30°C equal? Figure out by counting the number of short lines between 20°C and 30°C.

b. Give the temperature of the thermometer in the picture at right.

c. Round the following temperature values to the nearest 0.5°C: 23.1°C, 29.8°C, 30.0°C, 31.6°C, 31.4°C.

d. Water at 0°C and 100°C has different properties. Describe what water looks like at these temperatures.

e. What will happen to the level of the alcohol if you hold the thermometer by the bulb?

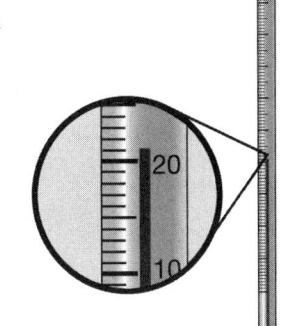

139

3. Reading the temperature of water in a beaker

An immersion thermometer must be placed in liquid up to the solid line on the thermometer (at least 2 1/2 inches of liquid). Wait about 3 minutes for the temperature of the thermometer to equal the temperature of the liquid. Record the temperature to the nearest 0.5°C when the level stops moving.

1. Place the thermometer in the beaker. Check to make sure that the water level is above the solid line on the thermometer.

2. Wait until the alcohol level stops moving (about 3 minutes). Record the temperature to the nearest 0.5°C.

4. Reading the temperature of warm water in a beaker

A warm liquid will cool to room temperature. For a warm liquid, record the warmest temperature you observe before the temperature begins to decrease.

1. Repeat the procedure above with a beaker of warm (not boiling) water.

2. Take temperature readings every 30 seconds. Record the warmest temperature you observe.

5. Reading the temperature of ice water in a beaker

When a large amount of ice is added to water, the temperature of the water will drop until the ice and water are the same temperature. After the ice has melted, the cold water will warm to room temperature.

1. Repeat the procedure above with a beaker of ice and water.

2. Take temperature readings every 30 seconds. Record the coldest temperature you observe.

Calculating Volume

How do you find the volume of a three-dimensional shape?

Volume is the amount of space an object takes up. If you know the dimensions of a solid object, you can find the object's volume. A two-dimensional shape has length and width. A three-dimensional object has length, width, and height. This investigation will give you practice finding volume for different solid objects.

Materials
- Pencil
- Calculator

1. Calculating volume of a cube

A cube is a geometric solid that has length, width, and height. If you measure the sides of a cube, you will find that all the edges have the same measurement. The volume of a cube is found by multiplying length times width times height. In the cube pictured at right, each side is 4 centimeters, so the problem looks like this:

$$V = l \times w \times h$$

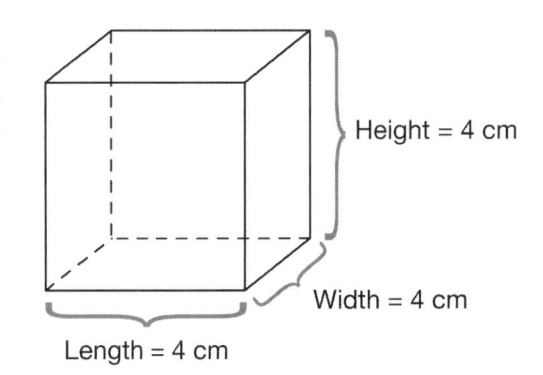

Height = 4 cm
Width = 4 cm
Length = 4 cm

Example:

Volume = 4 centimeters × 4 centimeters × 4 centimeters = 64 centimeters^3

2. Stop and think

a. What are the units for volume in the example above?

b. In the example above, if the edge of the cube is 4 inches, what will the volume be? Give the units.

c. How is finding volume different from finding area?

d. If you had cubes with a length of 1 centimeter, how many would you need to build the cube in the picture above?

3. Calculating volume of a rectangular prism

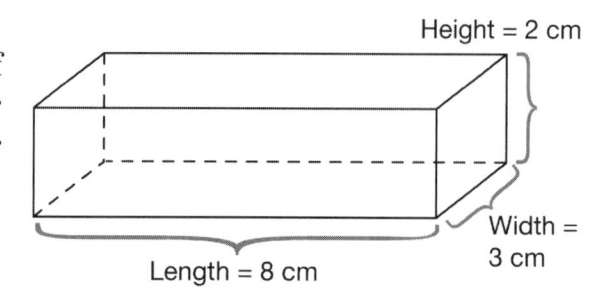

Rectangular prisms are like cubes, except not all of the sides are equal. A shoebox is a rectangular prism. You can find the volume of a rectangular prism using the same formula given above.
$$V = l \times w \times h$$
Or, multiply the area of the base times the height.

1. Find the area of the base for the rectangular prism pictured above.
2. Multiply the area of the base times the height. Record the volume of the rectangular prism.
3. PRACTICE: Find the volume of a rectangular prism with a height 6 cm, length 5 cm, and width 3 cm. Be sure to include the units in your answer.

4. Calculating volume for a triangular prism

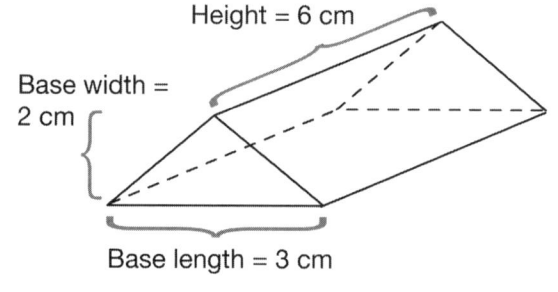

Triangular prisms have three sides and two triangular bases. The volume of the triangular prism is found by multiplying the area of the base times the height. The base is a triangle.

1. Find the area of the base by solving for the area of the triangle: $B = \frac{1}{2} \times l \times w$.
2. Find the volume by multiplying the area of the base times the height of the prism: $V = B \times h$. Record the volume of the triangular prism shown above.
3. PRACTICE: Find the volume of a triangular prism with a height 10 cm, triangular base width 4 cm, and triangular base length 5 cm.

5. Calculating volume for a cylinder

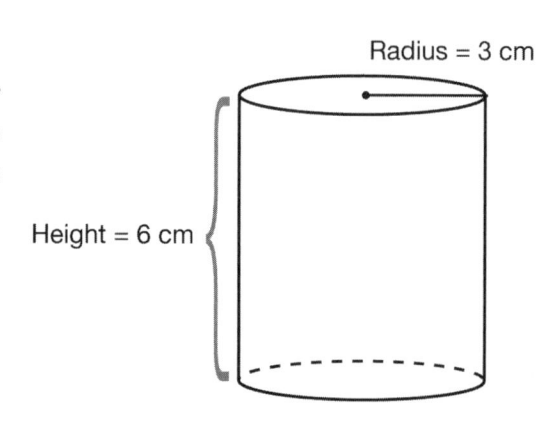

A soup can is a cylinder. A cylinder has two circular bases and a round surface. The volume of the cylinder is found by multiplying the area of the base times the height. The base is a circle.

1. Find the area of the base by solving for the area of a circle: $A = \pi \times r^2$.
2. Find the volume by multiplying the area of the base times the height of the cylinder: $V = A \times h$. Record the volume of the cylinder.
3. PRACTICE: Find the volume of a cylinder with height 8 cm and radius 4 cm.

6. Calculating volume for a cone

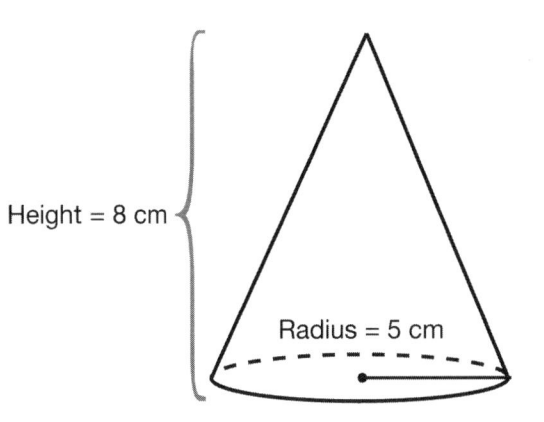

An ice cream cone really is a cone! A cone has height and a circular base. The volume of the cone is found by multiplying $1/2$ times the area of the base times the height.

1. Find the area of the base by solving for the area of a circle: $A = \pi \times r^2$.
2. Find the volume by multiplying 1/2 times the area of the base times the height: $V = 1/2 \times A \times h$. Record the volume of the cone.
3. PRACTICE: Find the volume of the cone with height 8 cm and radius 4 cm. Contrast your answer with the volume you found for the cylinder with the same dimensions. What is the difference in volume? Does this make sense?

Height = 8 cm
Radius = 5 cm

7. Calculating the volume for a rectangular pyramid

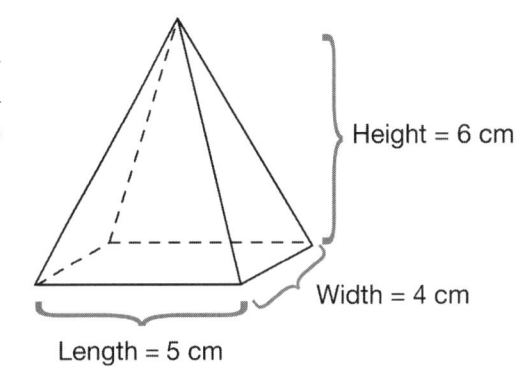

A pyramid looks like a cone. It has height and a rectangular base. The volume of the rectangular pyramid is found by multiplying $1/3$ times the area of the base times the height.

1. Find the area of the base by multiplying the length times the width: $A = l \times w$.
2. Find the volume by multiplying $1/3$ times the area of the base times the height: $V = 1/3 \times A \times h$. Record the volume of the rectangular pyramid shown above.
3. PRACTICE: Find the volume of a rectangular pyramid with height 10 cm, width 4 cm, and length 5 cm.
4. EXTRA CHALLENGE: If a rectangular pyramid had a height of 8 cm and a width of 4 cm, what length would it need to have to give the same volume as the cone in practice question 4 above?

Height = 6 cm
Width = 4 cm
Length = 5 cm

143

8. Calculating volume for a triangular pyramid

A triangular pyramid is like a rectangular pyramid, but its base is a triangle. Find the area of the base first. Then calculate the volume by multiplying $1/3$ times the area of the base times the height.

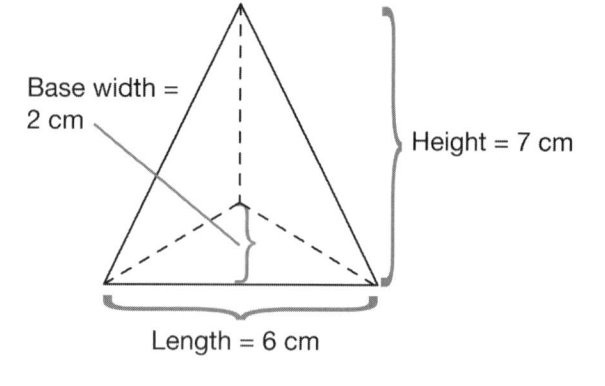

1. Find the area of the base by solving for the area of a triangle: $B = 1/2 \times l \times w$.
2. Find the volume by multiplying $1/3$ times the area of the base times the height: $V = 1/3 \times A \times h$. Find the volume of the triangular pyramid shown above.
3. PRACTICE: Find the volume of a triangular pyramid with height 10 cm, width 6 cm, and length 5 cm.

9. Calculating volume for a sphere

To find the volume of a sphere, you only need to know one dimension about the sphere, its radius.

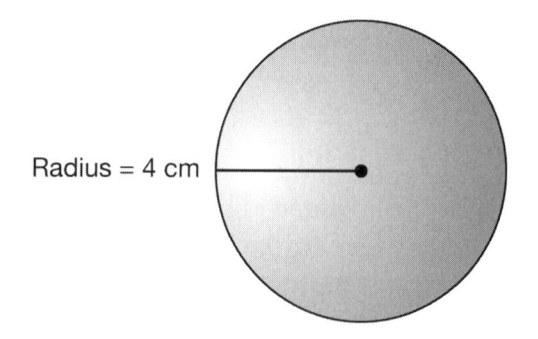

1. The volume of a sphere: $V = 4/3 \pi r^3$. Find the volume for the sphere shown at right.
2. PRACTICE: Find the volume of a sphere with radius 2 cm.
3. EXTRA CHALLENGE: Find the volume of a sphere with diameter 10 cm.

Measuring Volume

How do you find the volume of an irregular object?

It's easy to find the volume of a shoebox or a basketball. You just take a few measurements, plug the numbers into a math formula, and you have figured it out. But what if you want to find the volume of a bumpy rock, or an acorn, or a house key? There aren't any simple math formulas to help you out. However, there's an easy way to find the volume of an irregular object, as long the object is waterproof!

Materials
- Displacement tank
- Water source
- Disposable cup
- Beaker
- Graduated cylinder
- Sponges or paper towel
- Object to be measured

1. Setting up the displacement tank

Set the displacement tank on a level surface. Place a disposable cup under the tank's spout. Carefully fill the tank until the water begins to drip out of the spout. When the water stops flowing, discard the water collected in the disposable cup. Set the cup aside and place a beaker under the spout.

2. Stop and think

a. What do you think will happen when you place an object into the tank?

b. Which object would cause more water to come out of the spout, an acorn or a fist-sized rock?

c. Why are we interested in how much water comes out of the spout?

d. Explain how the displacement tank measures volume.

3. Measuring volume with the displacement tank

1. Gently place a waterproof object into the displacement tank. It is important to avoid splashing the water or creating a wave that causes extra water to flow out of the spout. It may take a little practice to master this step.

2. When the water stops flowing out of the spout, it can be poured from the beaker into a graduated cylinder for precise measurement. The volume of the water displaced is equal to the object's volume.
   Note: Occasionally, when a small object is placed in the tank, no water will flow out. This happens because an air bubble has formed in the spout. Simply tap the spout with a pencil to release the air bubble.

3. If you wish to measure the volume of another object, don't forget to refill the tank with water first.

Measuring Mass with a Triple Beam Balance

How do you find the mass of an object?

Why can't you use a bathroom scale to measure the mass of a paper clip? You could if you were finding the mass of a lot of them at one time! To find the mass of objects less than a kilogram, you will need to use the triple beam balance.

Materials

- Triple beam balance
- Small objects
- Mass set (optional)
- Beaker

1. Parts of the triple beam balance

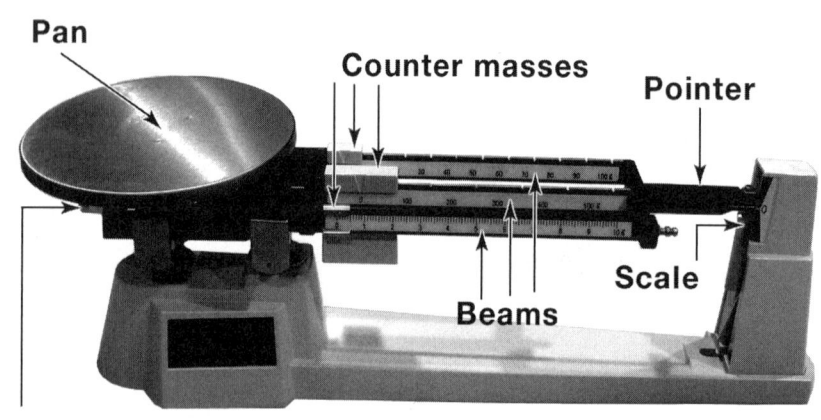

2. Setting up and zeroing the balance

The triple beam balance works like a see-saw. When the mass of your object is perfectly balanced by the counter masses on the beam, the pointer will rest at 0. Add up the readings on the three beams to find the mass of your object. The unit of measure for this triple beam balance is grams.

1. Place the balance on a level surface.

2. Clean any objects or dust off the pan.

3. Move all counter masses to 0. The pointer should rest at 0. Use the adjustment screw to adjust the pointer to 0, if necessary. When the pointer rests at 0 with no objects on the pan, the balance is said to be zeroed.

3. Finding a known mass

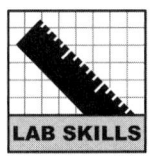

You can check that the triple beam balance is working correctly by using a mass set. Your teacher will provide the correct mass value for these objects.

1. After zeroing the balance, place an object with a known mass on the pan.

2. Move the counter masses to the right, one at a time, from largest to smallest. When the pointer is resting at 0, the numbers under the three counter masses should add up to the known mass.

3. If the pointer is above or below 0, recheck the balance setup. Recheck the position of the counter masses. Counter masses must be properly seated in a groove. Check with your teacher to make sure you are getting the correct mass before finding the mass an unknown object.

4. Finding the mass of an unknown object

1. After zeroing the balance, place an object with an unknown mass on the pan. Do not place hot objects or chemicals directly on the pan

2. Move the largest counter mass first. Place it in the first notch after 0. Wait until the pointer stops moving. If the pointer is above 0, move the counter mass to the next notch. Continue to move the counter mass to the right, one notch at a time, until the pointer is slightly above 0. Go to step 3. If the pointer is below 0, move the counter mass back one notch. When the pointer rests at 0, you do not need to move any more counter masses.

3. Move the next largest counter mass from 0 to the first notch. Watch to see where the pointer rests. If it rests above 0, move the counter mass to the next notch. Repeat until the point rests at 0, or slightly above. If the pointer is slightly above 0, go to step 4.

4. Move the smallest counter mass from 0 to the position on the beam where the pointer rests at 0.

5. Add the masses from the three beams to get the mass of the unknown object. You should be able to record a number for the hundreds place, the tens place, the ones place, the tenths place, and the hundredths place. The hundredths place can be read to 0.00 or 0.05. You may have zeros in your answer.

5. Reading the balance correctly

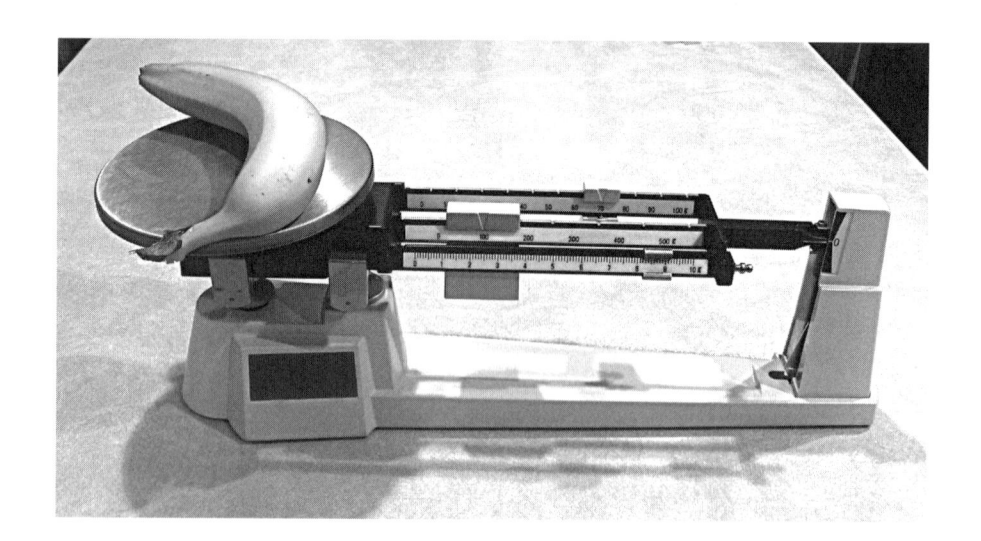

Look at the picture above. To find the mass of the object, locate the counter mass on each beam. Read the numbers directly below each counter mass. You can read the smallest mass to 0.05 grams. Write down the three numbers. Add them together. Report your answer in grams. Does your answer agree with others? If not, check your mass values from each beam to find your mistake.

6. Finding the mass of a substance in a container

To measure the mass of a liquid or powder you will need an empty container on the pan to hold the sample. You must find the mass of the empty container first. After you place the liquid or powder in the container and find the total mass, you can subtract the container's mass from the total to find the mass of the substance.

1. After zeroing the balance, place a beaker on the pan.

2. Follow directions for finding the mass of an unknown liquid or powder. Record the mass of the beaker.

3. Place the liquid or powder in the beaker.

4. Move the counter masses to the right, largest to smallest, to find the total mass.

5. Subtract the beaker's mass from the total mass. This is the mass of your unknown substance in grams.

Using a Compound Microscope

How do you use a compound microscope to see objects?

Have you ever used a magnifying glass? Objects under the magnifying glass look larger than real life. A compound microscope is like a magnifying glass that uses more than one lens to magnify small objects. In this investigation, you will become familiar with the parts of the compound microscope. Then you will examine a specimen with the microscope and practice using different levels of magnification.

Materials

- Compound microscope
- Power supply
- Clean glass slides
- Permanent markers
- Prepared specimens on glass slides
- Lens paper

1. Identifying the parts of the microscope

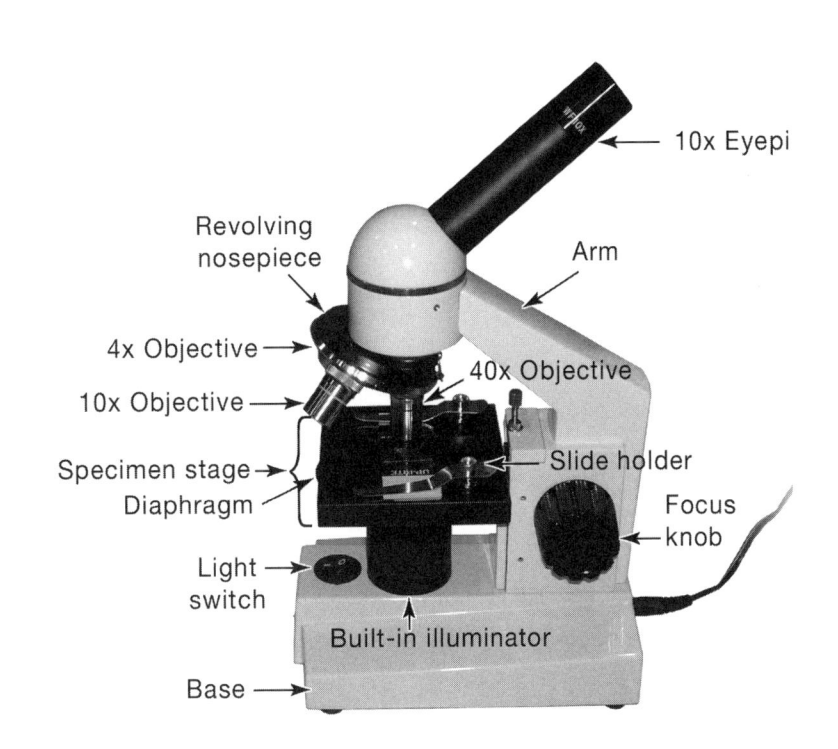

Look at the picture above. Each part of the microscope is labeled. The major parts of the microscope are the light source, the specimen stage, the eyepiece, and the objective lenses. You will find a description of the microscope parts at the end of this write-up.

2. Care of the microscope

Microscopes are expensive pieces of equipment containing glass parts that can break or scratch easily. Never touch the glass with your fingers.

1. Always carry the microscope with two hands. Hold the arm of the microscope with one hand and support the base with the other hand. Never turn the microscope upside down; the eyepiece could fall out.

2. Place the microscope on a level surface.

3. Check to make sure the battery-operated built-in illuminator is working. The microscope comes with a battery charger that can be plugged into an electrical outlet, if needed.

4. Without removing them, inspect the objectives, eyepiece, and illuminator for dust. If necessary, wipe the glass surfaces with lens paper. Store the microscope under a dust cover to keep it clean.

5. When you are finished using the microscope, switch to the lowest power objective (4x), lower the stage, switch off the power, cover the microscope with a dust cover, and return the microscope to its storage area.

3. Setting up the microscope

The microscope allows you to look in the eyepiece and see an image of the object on the stage. There are some differences between the image and the object. By following these directions, you will see how the image differs from the object.

1. With your teacher's permission, write a small letter "e" on a glass slide with a permanent marker.

2. Turn on the built-in illuminator.

3. Place the slide on the stage with the letter "e" facing you. Secure the slide under the slide holders.

4. Lower the stage until it is about halfway between the built-in illuminator and the objective.

5. Move the revolving nosepiece slowly until the 4x objective clicks into place and directly above the object.

6. Look into the eyepiece. Adjust the focus knob until the image appears in focus.

7. Carefully move the slide until the image is centered. Focus again if necessary. What do you notice about the direction of the letter "e"?

8. What happens to the image when you move the slide to the right? To the left? Forward? Backward? Specimens appear reversed when you look at them under the microscope because a mirror is used inside the microscope to direct the light rays to the eyepiece.

4. Adjusting the microscope

The microscope is adjustable so you can look at different types of samples. The following directions explain the adjustments you can make to get the best images possible.

1. Use the focus knob to move the stage up and down to see your specimen more clearly. The stage should be lowered when you are putting a slide on or off the stage.

2. Change the objective you are using by rotating the revolving nosepiece until a new objective clicks into place directly above the slide. An objective with a larger number increases the size of the image. Each objective has its magnifying power stamped on it.

   a. Find the 4x, the 10x, and the 40x objectives.

   b. Always start with the 4x objective directly above the slide when looking at a sample for the first time. View the sample through the eyepiece and bring it into focus.

   c. Center the object by moving the slide with your hand before changing the objective.

   d. Objectives are said to be parfocal if you can change from one objective to another without having to refocus the image very much. See if your image appears in focus when you switch to the 10x objective.

   e. When you use a 10x objective and a 10x eyepiece, the object appears one hundred times larger than its actual size. This number is called the total magnification. To solve for the total magnification, multiply the number on the objective times the number on the eyepiece. Solve for the total magnification for a 4x objective and a 10x eyepiece. Repeat the calculation using the 40x objective and the 10x eyepiece.

   f. After increasing the magnification, you will notice two things about the image. The image may appear darker, and the field of view becomes smaller. Objects near the edge of the image may disappear when you switch to a higher magnification. If your specimen is no longer visible, center the image at low magnification before switching to a higher magnification. These two effects are the normal result of switching to a higher magnification.

   g. Rotate the nosepiece slowly to move the 40x objective into place. Watch the microscope from the side while you do this. The 40x objective is very close to the stage when it is used correctly. The 40x objective is spring-loaded and retractable to prevent damage to the objective and slide. View your specimen through the eyepiece and adjust the focus if necessary.

   h. At higher magnification the amount of light entering the objective decreases. To increase the amount of light passing through the specimen, rotate the diaphragm under the stage. Select one of six different size holes to control the amount of light passing through.

151

5. Examining specimens under a microscope

Your teacher will provide you with prepared specimens on slides. In some cases, a stain has been used to allow you to see the specimen better. Follow the steps below to set up the microscope, and record your observations.

1. Lower the stage until it is about halfway between the built-in illuminator and the objective.

2. Turn on the built-in illuminator.

3. Place the slide on the stage and secure it with the slide holders.

4. Move the revolving nosepiece slowly until the 4x objective clicks into place and is directly above the object.

5. Look into the eyepiece. Bring the slide into focus using the focus knob.

6. Make a detailed sketch of what you see in the space below. Record your specimen name and your observations

7. Use the 10x objective and repeat steps 5 and 6.

8. Use the 40x objective and repeat steps 5 and 6.

Specimen: _____
(Low power)

Observations:
_____
_____

Specimen: _____
(Medium power)

Observations:
_____
_____

Specimen: _____
(High power)

Observations:
_____
_____

Glossary of microscope parts

- **10x Eyepiece**: One (monocular) or two (binocular) lenses that you look through to see the image
- **Built-in pointer:** A pointer inside the eyepiece to help you center your image
- **Arm:** Supports the upper half of the microscope
- **4x Objective:** Low power scanning objective
- **10x Objective:** Medium power objective
- **40x Objective:** High power objective
- **Revolving nosepiece:** Can be rotated to change objectives
- **Specimen stage:** The platform for holding the slide
- **Focus knob:** The knob that adjusts the height of the stage
- **Specimen:** Sample to be observed on the slide
- **Slide:** Glass support for specimens
- **Cover slip:** Glass cover for specimens
- **Light switch:** Switch to turn on built-in illuminator
- **Built-in illuminator:** Light source required to view specimens
- **Base:** Supports entire microscope
- **Power cord:** For recharging batteries to illuminator
- **Slide holder:** Clips to hold slide to stage
- **Diaphragm:** Adjusts amount of light entering specimen

Recording Observations in the Lab

How do you record valid observations for an experiment in the lab?

When you conduct an experiment, you make important observations. You and others will use these observations to test a hypothesis. In order for an experiment to be valid, the evidence you collect must be objective and repeatable. This investigation will give you practice in making and recording good observations.

Materials
- Paper
- Pencil
- Calculator
- Ruler

1. Making valid observations

Valid scientific observations are objective and repeatable. Scientific observations are limited to one's senses and the equipment used to make these observations. An objective observation means that the observer describes only what happened. The observer uses data, words, and pictures to describe the observations as thoroughly and precisely as possible. An experiment is repeatable if other scientists can perform it and achieve the same result. The following exercise gives you practice identifying good scientific observations.

2. Exercise 1

1. **Which observation is the most objective? Circle the correct letter.**
    a. My frog died after 3 days in the aquarium. I miss him.
    b. The frog died after 3 days in the aquarium. We will test the temperature and water conditions to find out why.
    c. Frogs tend to die in captivity. Ours did after three days.

2. **Which observation is the most descriptive? Circle the correct letter.**
    a. After weighing 3.000 grams of sodium bicarbonate into an Erlenmeyer flask, we slowly added 50.0 milliliters of vinegar. The contents of the flask began to bubble.
    b. We weighed the powder into a glass container. We added acid. It bubbled a lot.
    c. We saw a fizzy reaction.

3. Which experiment has enough detail to repeat? Circle the correct letter.

   a. Each student took a swab culture from his or her teeth. The swab was streaked onto nutrient agar plates and incubated at 37°C.

   b. Each student received a nutrient agar plate and a swab. Each student performed a swab culture of his or her teeth. The swab was streaked onto the agar plate. The plates were stored face down in the 37°C incubator and checked daily for growth. After 48 hours the plates were removed from the incubator, and each student recorded his or her results.

   c. Each student received a nutrient agar plate and a swab. Each student performed a swab culture of his or her teeth. The swab was streaked onto the agar plate. The plates were stored face down in the 37°C incubator and checked daily for growth. After 48 hours the plates were removed from the incubator, and each student counted the number of colonies present on the surface of the agar.

3. Recording valid observations

As a part of your investigations you will be asked to record observations on a skill sheet or in the results section of a lab report. There are different ways to show your observations. Here are some examples:

1. **Short description:** Use descriptive words to explain what you did or saw. Write complete sentences. Give as much detail as possible about the experiment. Try to answer the following questions: What? Where? When? Why? and How?

2. **Tables:** Tables are a good way to display the data you have collected. Later, the data can be plotted on a graph. Be sure to include a title for the table, labels for the sets of data, and units for the values. Check values to make sure you have the correct number of significant figures.

Table I: U.S. penny mass by year

Year manufactured	1977	1978	1979	1980	1981	1982	1983	1984	1985
Mass (grams)	3.0845	3.0921	3.0689	2.9915	3.0023	2.5188	2.5042	2.4883	2.5230

3. **Graphs and charts:** A graph or chart is a picture of your data. There are different kinds of graphs and charts: line graphs, trend charts, bar graphs, and pie graphs, for example. A line graph is shown below.

   Label the important parts of your graph. Give your graph a title. The *x*-axis and *y*-axis should have labels for the data, the unit values, and the number range on the graph.

   The line graph in the example has a straight line through the data. Sometimes data does not fit a straight line. Often scientists will plot data first in a trend chart to see how the data looks. Check with your instructor if you are unsure how to display your data.

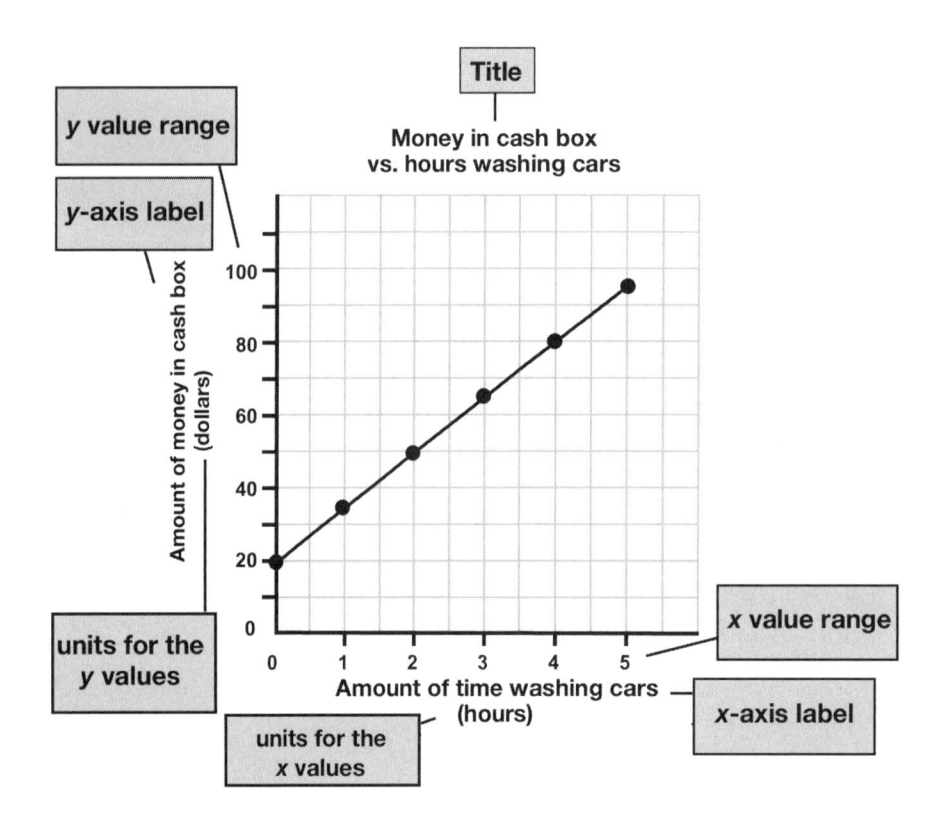

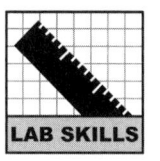

4. **Drawings:** Sometimes you will record observations by drawing a sketch of what you see. The example below was observed under a microscope.

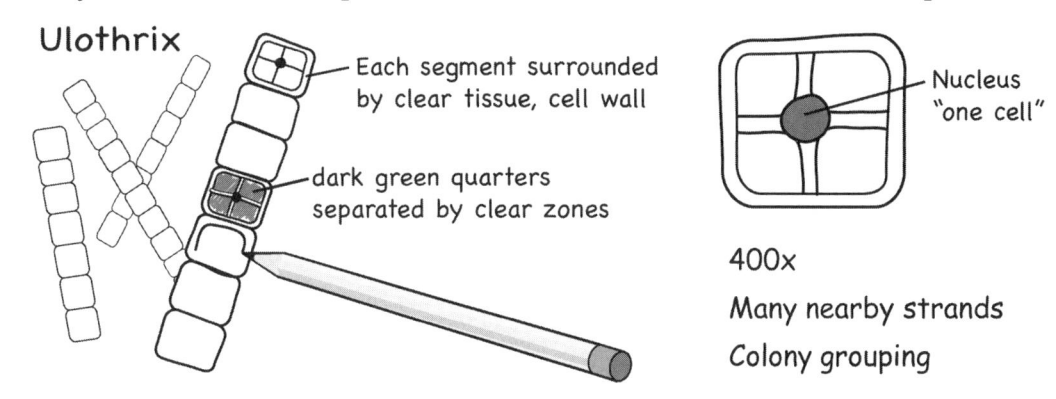

Give the name of the specimen. Draw enough detail to make the sketch look realistic. Use color, when possible. Identify parts of the object you were asked to observe. Provide the magnification or size of the image.

Exercise 2: Practice recording valid observations

A lab report form has been given to you by your instructor. This exercise gives you a chance to read through an experiment and fill in information in the appropriate sections of the lab report form. Use this opportunity to practice writing and graphing scientific observations. Then answer the following questions about the experiment.

A student notices that when he presses several pennies in a pressed penny machine, his brand new penny has some copper color missing and he can see silver-like material underneath. He wonders, "Are some pennies made differently than others?" The student has a theory that not all U.S. pennies are made the same. He thinks that if pennies are made differently now, he might be able to find out when the change occurred. He decides to collect a U.S. penny for each year from 1977 to the present, record the date, and take its mass. The student records the data in a table and creates a graph plotting U.S. penny mass vs. year. Below is a table of some of his data:

Table 2: U.S. penny mass by year

Year manufactured	1977	1978	1979	1980	1981	1982	1983	1984	1985
Mass (grams)	3.0845	3.0921	3.0689	2.9915	3.0023	2.5188	2.5042	2.4883	2.5230

Stop and think

a. What observation did the student make before he began his experiment?

b. How did the student display his observations?

c. In what section of the lab report did you show observations?

d. What method did you use to display the observations? Explain why you chose this method.

Measuring Biodiversity

How do you measure the biodiversity of an ecosystem?

<u>Biodiversity</u> refers to the measure of the variety and number of organisms that live in an area. To date, 1.8 million species have been identified. Although this is a big number, scientists believe that there are actually 5 to 100 million species on Earth. There are many species yet to be identified. Microorganisms are the least known. This means that there's a lot of work to do. Maybe you will be the next person to identify a new species! Biodiversity is also a measure of the health of an ecosystem. In general, the greater the biodiversity of an ecosystem, the healthier it is. In this investigation, you will work on a team to measure the biodiversity in two or more ponds by looking at samples of pond water. You will compile class data to get average values of biodiversity for each pond.

Materials

- Glass jar
- Pond or creek water
- Hay or dry grass
- Milk
- Yeast
- Depression slides
- Cover slips
- Corn syrup
- Light microscope
- Dropper
- Pond organisms chart

1 Setting up

1. Select two or more ponds in your area to study.
2. Write a short report that details what you know about each pond. For example, identify the location of each pond, describe the surroundings of the ponds, and state the apparent health of the ponds.

a. Develop a hypothesis that states which pond in your study will have the highest biodiversity. Write a paragraph that justifies your hypothesis.

2 Collecting data

For this investigation, you will measure biodiversity by measuring species richness. <u>Species richness</u> is the total number of species in a system.

1. Follow the instructions for Investigation 9B (Investigating Pond Water) to set up hay infusions for each of the ponds and to record your observations. Your teacher may have already prepared the samples for you.
2. For this investigation, examine a total of three samples from each hay infusion. Take a sample from each zone of the jar—the bottom, middle, and top (shown at right).
3. Prepare slides for observation from each zone as you did in Investigation 9B.

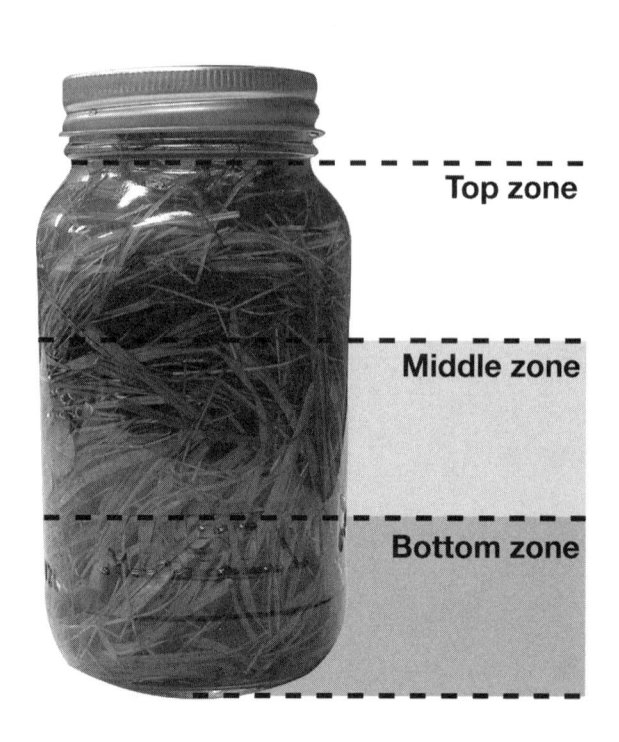

Investigation

EXPLORE further

4. Make sketches of each species you observe. Add your sketches to Table 1. Use separate paper if you need more space. Identify species by comparing sketches you made of the organisms to photos provided by your teacher. Use other resources such as books or the Internet if needed. You may also work with other teams to identify organisms. List the name of each species in Table 1. Note: If you cannot find the correct name for an organism, give it a temporary name.
5. Count the total number of species you observe in your samples from each jar. Record this data in Table 2.
6. Compile your team's data with the data gathered by other teams in your class. In Table 2, record the total number of species listed for each pond and the average values for species richness for each pond.

Table 1: Species names and sketches

Pond sample #1		Pond sample #2		Pond sample #3	
Species	Sketch	Species	Sketch	Species	Sketch

Table 2: Species richness

	Pond #1	Pond #2	Pond #3
Species richness (your team's data)			
Total number of species (all teams' data)			
Average species richness (Total number of species/total number of jars)			

159

3. Analyzing the data

a. Which pond had the highest species richness?

b. Why might it make sense to get an average value for species richness versus using the total number of species as your species richness value?

c. Do you feel your species richness value is accurate? Why or why not?

d. How could you improve the accuracy of your species richness value?

4. Evaluating your results

a. Based on the data you collected, was your hypothesis correct?

b. High species richness and biodiversity are associated with healthy ecosystems. Does your data support this statement? Why or why not?

c. Write a lab report that summarizes your work, findings, and conclusions.

5. Applying your knowledge

a. Were there any organisms that you could not identify? Describe a process that you might use to identify these organisms.

b. Microorganisms are the least known organisms in terms their level of biodiversity on Earth. Why do you think this is the case?

c. Scientific evidence suggests that most of the species that have ever lived on Earth have gone extinct and that today's level of biodiversity took a very long time to achieve. The real diversification of life began about 600 million years ago. Check out Figure 14.15 in your text, and read section 14.1 to find out what Earth was like 600 million years ago. What began to happen then? What was the name of this era?

d. In Chapter 14 you learned about _mass extinctions_ (periods of large-scale extinction). One mass extinction happened 65 million years ago. At this time (the end of the Cretaceous Period), 60 to 70 percent of all plant and animal species vanished. This included the disappearance of Earth's large vertebrates (such as the dinosaurs) and most of the oceans' plankton. Earth's biodiversity was greatly diminished. Eventually, however, biodiversity returned to pre-extinction levels. Why was this mass extinction advantageous for present-day organisms?

e. What happens to people when ecosystems experience a loss of biodiversity? One result is that resources people depend on—such as food, energy, and medicines—are lost. Less species rich ecosystems are also not as able to serve as places where nutrients get recycled and where water gets filtered. Research this issue and write a position paper that explains why it is important to protect Earth's biodiversity. Include at least three examples that support your position.

Investigation

Innovation and Recycling

How can the engineering cycle be used to improve a product so that it can be better recycled?

Have you ever thrown something in the trash and regretted that you could not easily recycle it? In this exploration, you will use the engineering cycle to improve the recyclability of a product.

Materials

- Various materials that can be used to build a prototype
- Logbook or journal for notes, sketches, and data

1. Recycling research

The questions below will guide you as you research recycling and the trends that are taking place in the U.S. today. Use the Internet or local library or visit a recycling facility to conduct your research. Organize your answers into an essay or a poster.

a. What materials can be recycled, and how are they sorted at a recycling facility?

b. What happens to the materials after they are sorted?

c. What materials cannot be recycled at recycling facilities? What happens to those materials?

d. What does the phrase "zero waste" mean?

2. The engineering cycle

In this investigation, you will first select a product. Next, you will be innovative and redesign it so that it is easier to recycle or reuse. Innovation is the process of making improvements to an existing product. Ideally, your redesigned product will fulfill the criteria of creating "zero waste." This means that it can be completely reused or recycled and that zero or very little material from the product or its packaging enters a landfill. Use the engineering cycle to work with the product you select.

The parts of the engineering cycle are

- **Identify a Need:** Define the problem and set the goal of the project. Identify the constraints and variables.

- **Design:** Brainstorm to create a list of ideas for the project. Be creative. It is okay if some ideas seem far-fetched. Do some research to refine your ideas. Then, select the best ideas. Use the ideas to create a design. Choose materials, make drawings, and decide what you will build.

- **Prototype:** Follow your design to build a prototype. Keep a record of any difficulties you run into while building.

- **Test:** Find out if your prototype works. Make notes about its strengths and weaknesses.

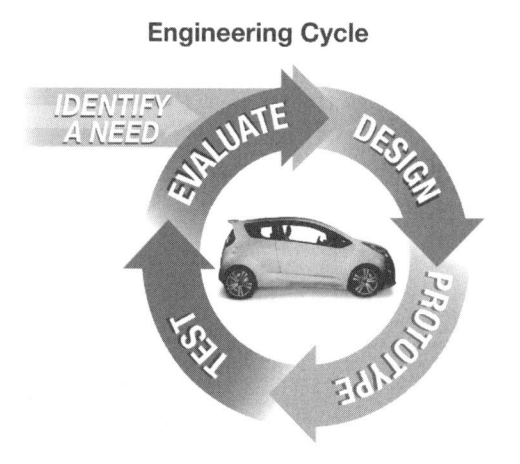

Engineering Cycle

- **Evaluate:** Once the prototype is tested, evaluate the design and brainstorm new ideas for improving it. The engineering cycle is repeated as many times as needed to create and improve products.

3 An example of the engineering cycle in action

Here is an example of the engineering cycle in action. The chosen product to innovate is the toothbrush.

Step	Example for redesigning a toothbrush
Identify the need	The problem: About 22 million kg of toothbrushes enter U.S. landfills each year. Goal: Redesign the toothbrush to make it biodegradable. Variables: Materials, size of brush, how long the toothbrush lasts Criteria: Effective in cleaning teeth, produces zero waste Constraints: Light-weight, inexpensive
Design	The entire toothbrush will be made of biodegradable material. The brush will be an attachment that can be replaced.
Prototype	Used biodegradable plastic cups to build a toothbrush handle that had a slot for inserting the removable, replaceable brush. The brush easily pops into place and can be popped off.
Test	Made a few prototypes and tried them out. The handle broke easily in every test.
Evaluate	The handle needs to be stronger and more rigid.

Recycling note: Various companies sell recyclable toothbrushes. Search online to discover the various ways the toothbrush has been redesigned, and purchase your favorite innovation.

Investigation

4. Using the engineering cycle

Keep a logbook to document how you follow the engineering cycle with the product you select. Your logbook should contain all of your notes, drawings, observations, and testing results. The guidelines below will help you work through the engineering cycle.

Identify a Need

> Make sure you are clear about the goals, criteria, and constraints of the project. These will affect the type of prototype you can build.
>
> Your innovated product
> - Must reduce the amount of waste going to landfills.
> - Must be recyclable or reusable.
>
> Note: Choose an inexpensive product so that you can easily purchase it for this investigation.

Design

> 1. Brainstorm ideas for a product to select. Then, select a product and purchase it.
> 2. Now, brainstorm ideas for how to improve this product. Make notes and sketches. List possible materials you can use to build a prototype.
> 3. Study your ideas. Select the best design. Make a drawing of your design. Include measurements and label the materials.

Prototype

> 1. Build your prototype.
> 2. Be sure to make notes in your logbook about any changes you make to the original design.

Test

> Test your prototype. Make notes about what worked well and what didn't work.

Evaluate

> Go back to the "design" step to improve your model. You may choose to use a completely different design, or you may want to think about ways to refine your current design. Keep going through the cycle until you are completely satisfied with your innovation.

5. Comparing and evaluating models

How do your classmates' products compare to yours? Make comments about what you like about each design and what could be changed to make it better. Here are some guidelines to help you rate each prototype.

1. Watch the product in action, or listen to a presentation by the team.
2. Rate the product 1 to 10 for design creativity, quality of construction, and recyclability.
3. Make comments about the strong points of the design and the ways it could be improved.

163

6. Reviewing your work

a. Write a short paragraph that explains why you choose your product.

b. In this exploration you practiced innovation. What is the difference between innovation and invention?

c. All products have a "life cycle." Life cycle analysis of a product takes into account the resources and raw materials used to make the product, the energy resources needed while the product was produced and transported to be sold, the energy resources used to sell and use the product, and all waste produced when the product is thrown away. Do a life cycle analysis of your innovated product. Explain why your innovation is an improvement upon the original.